Essentials for Deeper Understanding of Quantum Computing

Ryo Maezono

Essentials for Deeper Understanding of Quantum Computing

Springer

Ryo Maezono
School of Information Science
Japan Advanced Institute of Science
and Technology
Nomi, Ishikawa, Japan

ISBN 978-981-96-5645-5 ISBN 978-981-96-5646-2 (eBook)
https://doi.org/10.1007/978-981-96-5646-2

This Springer imprint is published by the registered company Springer Nature Singapore Pte Ltd.
The registered company address is: 152 Beach Road, #21-01/04 Gateway East, Singapore 189721, Singapore

Preface

"Are all introductory books on quantum computers this rigid and excessively difficult?"

This book is written with the goal of understanding quantum computers and quantum annealing. It avoids a formal scientific style and minimizes the elementary mathematics and physics surrounding quantum mechanics, presenting the material as a coherent narrative. When it comes to quantum computing, standard references such as Nielsen and Chuang's textbook provide a foundational chapter titled "Introduction to Quantum Mechanics," offering introductory knowledge. While some beginners may successfully navigate such content to achieve a basic understanding of quantum mechanics, there is also a significant portion of readers, including the author, who are less familiar with formal sciences and find that "it is not as easy to read as an 'introduction' should be."

Every time I write a paper or a book, I struggle with the careful selection of words, constantly concerned about whether "this might be misunderstood in that way" or "this could possibly be interpreted too broadly." The method humanity has invented to preserve "facts that need to be conveyed" accurately, and in a way that minimizes misunderstanding or misinterpretation, is what we call the "formal scientific method of description." Each "system of knowledge," such as physics or quantum mechanics, begins with hand-waving discussions, gradually becoming systematized through storytelling-like styles akin to those in natural science books, and ultimately evolves into the "formal scientific method of description." In recent years, introductory books on quantum mechanics have increasingly adopted styles that emphasize the formal scientific approach.

However, perhaps due in large part to the author's own limitations, I often find myself thinking, "Unless you've majored in mathematics, this is nearly impossible to read," or "Was it really supposed to be this difficult?" The formal scientific method of description is akin to a "preservation technique that allows content to be stored across time and space." Mathematics books are a prime example; they require the "careful soaking and rehydrating" equivalent of "rereading repeatedly to understand even a single section." For readers coming from fields without experience in the "rigorous training of formal sciences," the harsh reality may be that without exceptionally strong determination to study, it is quite challenging to work through "introductory quantum mechanics written in a formal scientific style."

At the beginning of books written in a formal scientific style, it is often argued that: "The traditional approach to quantum mechanics takes too much time. Lengthy explanations of experimental facts and historical context dampen students' motivation to learn. By adopting a formal scientific style that clarifies the structure of core principles and their consequences, readers can proceed without losing their desire to learn." However, for readers less familiar with formal sciences, it is precisely the "formal scientific style, which makes decoding even a single page a struggle," that can diminish their motivation to continue reading. There is undoubtedly a segment of readers who feel that books written in a "narrative style" are less discouraging and more engaging to read. With this in mind, I developed the content of this book, believing that such books should also have their place.[1]

For many years, I have been teaching a course aligned with the content of this book, which could be described as a "minimum physics course for information science students," within the Graduate School of Information Science. In our institution, which only offers graduate programs, students come from a variety of undergraduate backgrounds, and those familiar with formal sciences, such as students from information science or mathematics, are actually in the minority. As a result, the aim of providing a "curriculum that avoids the formal scientific style" seems to have been reasonably successful. The course is designed with quantum computers and quantum annealing as its ultimate goals, trimming down the content to include only what is necessary, and fitting into approximately 2 credits worth of material (14 sessions of 90-minute lectures). The content was structured to form a "single climbing route" to this goal. In a small class of about 20 students, the course is conducted entirely via chalkboard lectures. I adjust the pace based on each student's reactions, including casual discussions and providing more detailed explanations if someone seems confused. At this pace, the course slightly exceeds 14 sessions, requiring around 16 sessions. However, in a larger lecture hall setting, where individual student reactions are not monitored and PowerPoint slides are used to maintain the pace, the material could comfortably fit within the allotted 14 sessions.

With quantum computers and quantum annealing as the ultimate goals, I structured the content to ensure "a solid minimum foundation" that students can feel confident about. The basic concepts introduced to beginners were narrowed down to three key topics: "quantum mechanical superposition," "tunneling phenomena," and "spin degrees of freedom." Rather than presenting these tools in a top-down manner, the course is designed as a tutorial to provide a clear view of "what kind of ideas led to these topics and the pathways through which they emerged." To present these

[1] During my student days, I often felt that "physics books seem to become increasingly difficult in their writing style as time goes on." As a strategy, I would sometimes turn to "books written in the immediate postwar period, often in older Chinese characters," and I frequently came across "lively, narrative-style explanations" in them. These books greatly aided my understanding and were an invaluable resource for me.

topics naturally, knowledge of analytical mechanics and electromagnetism becomes necessary, and understanding these topics requires a corresponding reinforcement of basics in elementary mathematics.

Regarding the assumed level of readers, this book is aimed at students who may not confidently claim to "fully understand basics in elementary mathematics at the undergraduate level." From years of experience, I have identified that the primary reason these students often give up on mastering basics of mathematics lies in their unfamiliarity with "index operations." Therefore, the early chapters of this book provide particularly thorough explanations of this topic. Once this hurdle is cleared, students will find it much easier to follow mathematical operations in linear algebra, vector calculus, and the foundational physics that follows. By reducing the difficulties encountered in mathematical operations, the core idea that "linear algebra is a framework for representing mappings" becomes clearer. The central theme running through the topics covered in this book is "how the rules of description change when the coordinate system is transformed and how the quantities of interest adapt to these changes." Thus, the progression "from index operations to linear algebra" provides the foundational skills needed to understand the later chapters.

The ultimate goal of this book is quantum annealing, which serves as the culmination of the narrative. However, the foundational mathematics and analytical mechanics covered in the earlier chapters also provide fundamental skills that are beneficial for understanding other topics. These foundations are not only useful in material science, such as in electronic structure calculations, which were the original motivation for writing this book, but also in the field of information science. For example, the content connects to: machine learning (vector calculus $\rightarrow$ optimization mathematics), robotics (analytical mechanics $\rightarrow$ mechanical dynamics), and reinforcement learning (analytical mechanics $\rightarrow$ the background of the Bellman equation). Thus, the material presented here has been developed and taught to support a wide range of applications across these domains.

As described above, this book does not assume high-level mathematical analysis skills from its readers and provides detailed explanations. The content is accessible to readers such as those "not from a science background but with a solid understanding of high school mathematics" or high school students interested in mathematical sciences. To avoid losing sight of the overall structure, derivations and proofs that are easily found in other textbooks are delegated to the references cited. However, I have strived to present a clear narrative explaining "the logical flow that leads to these topics." Particular attention has been given to ensuring that readers outside the field can feel assured that they have "learned a proper framework," crafting the material into a coherent and understandable progression.

Guiding Principles of Writing

In Chap. 2, which covers foundational mathematics, a relatively large portion of the book is devoted to this topic. In my previous books, I adopted a similar structure, for example, allocating significant space to Linux operations in the textbook aiming at "cluster server construction and administration." My guiding principles in writing are: "Introduce the foundational topics thoroughly and carefully, without overwhelming readers with excessive advanced topics," and "Narrowing down the required elements necessary to understand the logical progression of subsequent topics." These principles have been carried over from my earlier works to publish textbooks.[2] This book is designed with the intention of being used in lectures. The content to be taught within the limited time of a lecture should focus on "carefully introducing the key concepts," while supporting students in acquiring the skills necessary for self-study on the details of advanced topics.

An alternative perspective emphasizes that "the role of a lecture is to adopt the standard textbook of the field as the Bible and cover its contents comprehensively." In fact, this approach to course design is more common, and as such practices are uncritically passed down, it has led to lecture formats where "nearly 100 slides are read aloud like a picture story show during a single 90-minute lecture".[3] Students subjected to such "read-aloud slide presentations," which could easily be replaced by pre-recorded automated narration, seem rather unfortunate.

However, when teaching small graduate-level classes while closely monitoring student reactions, I often notice situations like this: Even after relatively thoroughly covering topics such as the diagonalization process in linear algebra, I am often surprised to realize that students, while nodding and appearing engaged during the lecture, may not actually understand the concrete expression of basis orthogonality, $\mathbf{e}_1 \cdot \mathbf{e}_2 = 0$. Rushing to cover "advanced topics ahead" should be avoided. If the introductory sections are not taught with substantial care and thoroughness, students may end up spending more than half the lecture merely "copying scripture."

The idea that "those aspiring to engage in research at the graduate level must thoroughly digest the standard masterpieces of their field" is one I wholeheartedly agree with. However, this should be accomplished not in the lecture setting, but through careful and thorough independent study. If everyone could easily self-study from the standard works of the field, there would be no need for university lecture courses. The reason self-study is difficult is that students often do not grasp the core concepts or overarching structure. It is not that they fail to understand each operation line by line, but rather that they have no idea what the text is trying to convey as a whole. Over time, this lack of clarity erodes their motivation to continue studying independently. This is where the role of the teacher as a guide becomes important. I believe the purpose of a lecture is to provide "a well-structured introduction to these concepts and the overarching logic," offering a kind of "takeoff assistance" for students.

Advanced details can be independently studied according to each individual's specific objectives, as long as the basic concepts are firmly grasped. Based on this philosophy, this book devotes considerable depth and care to the sections covering foundational mathematics.

---

[2] This guiding principle is emphasized in the prefaces of all my previous textbooks.

[3] This is particularly noticeable in lectures delivered by Japanese instructors teaching in English.

Acknowledgments This book is based on the content of my lecture "Modeling of Dynamics," which was offered in the School of Information Science at the Japan Advanced Institute of Science and Technology (JAIST) for ten years, starting in the 2014 academic year. Through report assignments, I was able to identify the specific stages at which students encountered difficulties in understanding the material. In particular, the experience of identifying "which topics should be carefully developed for students who possess latent mathematical ability but have no prior background in physics" helped shape the unique characteristics of this book. This learning experience was made possible by the US-style graduate coursework system, a distinctive feature of JAIST. I would like to express my sincere gratitude for this.

Nomi, Japan
February 2025

Ryo Maezono

Contents

Chapter 1
What Is Unclear?

Regarding the question posed in this chapter's title, the common response is often, "Oh, it's the issue of quantum superposition that isn't clear. Even top-tier physicists find it unclear, so it's understandable". However, from my experience teaching beginners, I realize that the points where beginners feel unclear are not the same as the ones that top-tier physicists struggle with. They are much more elementary. In this chapter, I will identify these elementary points of confusion and set the goals of this book accordingly.[1]

Avoid Reading Too Closely

The content of this chapter is structured to resonate with readers who have studied these topics before but found them unclear. Readers who are relatively unfamiliar with the subject should take care **not to focus on reading every detail too closely**. This chapter is not structured in a way that requires full comprehension before proceeding to the next chapter. For complete beginners, unfamiliar notations may appear without prior definition, but these will be gradually introduced in the subsequent chapters, so there is no need for concern. Instead, simply **skim through** this chapter with the mindset of understanding the general flow and the intended objectives of the discussion that follows.

(continued)

[1] Few beginners can accurately explain "what they don't understand". As a result, "technical points of confusion for many beginners" are sometimes mistaken for "points that even top-tier physicists find unclear". To avoid confusing beginners with issues that top-tier physicists find unclear, many introductory books have been made more difficult for the majority of beginners struggling with technical points.

R. Maezono, *Essentials for Deeper Understanding of Quantum Computing*,
https://doi.org/10.1007/978-981-96-5646-2_1

It is important to note that "skimming through a text" is not an act of laziness or a negative habit, but rather a *skill that should be actively developed*. Even if only about half of the content is understood, it is crucial to persist in "picking up what is comprehensible and making it through to the end of the book". By doing so, one gains an overview of the chapter structure and becomes familiar with key terminology. Then, upon encountering the same topic again, understanding deepens gradually—like cutting through cardboard by progressively deepening the grooves with a utility knife. When the author transitioned from high school to university, this method of learning was initially difficult to adopt. Instead, following the study habits from high school, attempting to meticulously read every detail in a textbook often led to frustration and failure to progress beyond the second chapter. As a result, advanced research topics always seemed out of reach, a consequence of being unable to move beyond the stage of "not even making a groove in the cardboard".

1.1 How to Approach Quantum Mechanics

In courses related to quantum computing, it is often mentioned by physicists that "no one truly understands quantum mechanics" [1]. This statement can be quite discouraging for beginners wanting to learn quantum mechanics. To organize "what to learn about quantum mechanics" hierarchically:

1. How the governing principles were discovered,
2. Why those governing principles exist in the first place,
3. How to apply those governing principles,

provides a clearer perspective against the confusion. While it is true that "no one understands (2)", beginners should first aim to understand (1) and (3). Only after reaching a level where they understand these, they can empathize with the statement "no one understands (2)".

In tutorials on quantum computing [2], the question "Why does (3) work?" is often avoided, and instead, the axiomatic approach of "accept the operation as an axiom" is taken. On the other hand, in physics, chemistry, and materials science, the traditional approach spends time on (1). Prefaces of texts that adopt the axiomatic style often suggest that including (1) causes beginners to become confused and misunderstand, hence the axiomatic approach was chosen out of consideration. For beginners of quantum mechanics proficient in formal sciences [Beginners Type A],[2] the lecture course that treats (3) as an axiom can work very well. However, for those not sufficiently familiar with formal sciences [Beginners Type B], the overly symbolic nature of the axiomatic method can be overwhelming, making the material difficult to digest.[3]

Even those who have mastered (1) usually do not grasp it through a thoroughly logical flow. Often, their knowledge settles in after encountering the topic sporadically in undergraduate lectures and gradually becoming accustomed to it without feeling discomfort.

Seems Quite Different Books for the Same Topic...
A colleague in the mathematical sciences side once shared their surprise with me when trying to understand the broad concepts of quantum computing through various books labeled "Quantum Mechanics". They noted that there were two completely different approaches, making it seem like they weren't even discussing the same topic. This difference stems from whether the books primarily focused on (1) or (3).

In this chapter, we will first provide an overview of gate-based quantum computation and quantum annealing at the standard level of introductory books on quantum computation (Sect. 1.2). Following this overview, Sect. 1.3 will outline the technical points where beginners tend to get confused and set the goals for understanding these points in this book.

1.2 Introduction Overview at a Standard Granularity

In this section, we would identify and confirm the points of confusion like "I still don't get this". Therefore, beginners shouldn't worry if they don't fully understand everything right away; just read through to grasp the overall flow. The goal of this book is for you to eventually understand this section smoothly. It might be helpful to revisit this section periodically to check your understanding.

[2] Formal sciences are described in a structured format of axioms, definitions, theorems, and proofs. When attempting to articulate scientific theories, one would notice how easily one's wording can cause misunderstandings. The structured format of formal sciences aims to eliminate such 'ambiguity due to wording' as much as possible, being an ultimate method to accurately convey knowledge to future generations.

[3] A similar issue exists in teaching English to Japanese. Some learners can master English through listening and practical conversation without grammatical explanations. However, there is also a significant portion of learners who need explanations of grammar to understand the language, highlighting the diversity in learning approaches.

1.2.1 Quantum Gate-Based Computing

In digital computers, computational operations are implemented using the binary states $\{|0\rangle, |1\rangle\}$, which are realized exclusively [3]. In the quantum world, however, a **superposition state**

$$|\psi\rangle = c_0 \cdot |0\rangle + c_1 \cdot |1\rangle \tag{1.1}$$

is realized, where the coefficients $c_{0,1}$ are complex numbers (Sect. 5.2.3). This **quantum bit** $|\psi\rangle$ represents a state where "upon observation, one finds either $|j = 0, 1\rangle$ with probability $|c_j|^2$, but until observed, it exists as a simultaneous 'superposition' of both $|0\rangle$ and $|1\rangle$" (Sect. 5.2.3).

For example, using a combination of 3 quantum bits,[4]

$$\begin{aligned}
&|\psi_1\rangle \otimes |\psi_2\rangle \otimes |\psi_3\rangle \\
&= \left(c_0^{(1)} \cdot |0\rangle + c_1^{(1)} \cdot |1\rangle\right) \otimes \left(c_0^{(2)} \cdot |0\rangle + c_1^{(2)} \cdot |1\rangle\right) \otimes \left(c_0^{(3)} \cdot |0\rangle + c_1^{(3)} \cdot |1\rangle\right) \\
&= A_{000}|000\rangle + A_{001}|001\rangle + \cdots + A_{111}|111\rangle
\end{aligned}$$

then, until observation, all eight possible bit states from $|000\rangle$ to $|111\rangle$ exist simultaneously in a superposition.

Now, consider the initial state of the quantum bits being prepared as:

$$|\mathrm{Init}\rangle = A_{000}|000\rangle + A_{001}|001\rangle + \cdots + A_{111}|111\rangle$$

we consider how the output bits change when a gate operation $\hat{F}$ is applied to this initial state $\hat{F} \cdot |\mathrm{Init}\rangle$. The gate configuration is generally constructed to interfere the "initial state itself $\hat{I} \cdot |\mathrm{Init}\rangle$" with the "state where the phase of the coupling constants of the initial state is modulated"[5] as $\hat{G} \cdot |\mathrm{Init}\rangle$. Thus, $\hat{F}$ can be expressed as $\hat{F} = \left(\hat{I} + \hat{G}\right)$ as

$$\begin{aligned}
\hat{F} \cdot |\mathrm{Init}\rangle &= \left(\hat{I} + \hat{G}\right) \cdot |\mathrm{Init}\rangle \\
&= (A_{000} + G_{000})\,|000\rangle + (A_{001} + G_{001})\,|001\rangle \\
&\quad + \cdots + (A_{111} + G_{111})\,|111\rangle\ .
\end{aligned}$$

[4] Here, coefficients like $c_0^{(1)} c_1^{(2)} c_0^{(3)} = A_{010}$, and states like $|0\rangle \otimes |1\rangle \otimes |1\rangle = |011\rangle$ are abbreviated.

[5] Since $A_p = |A_p| \cdot e^{i\varphi}$ is a complex number, it has a phase φ. For example, $\varphi = \pi$ corresponds to a phase inversion (-1).

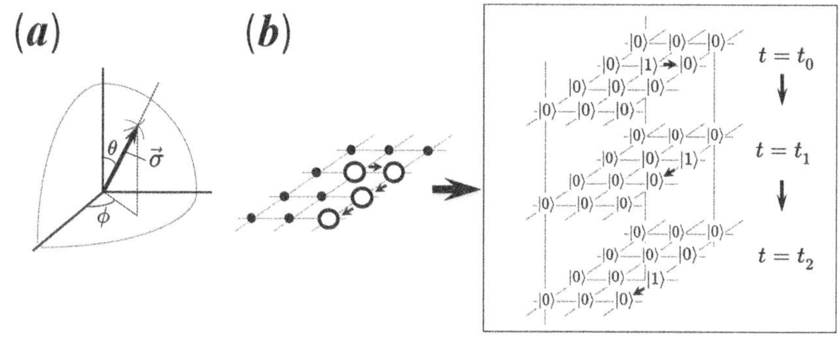

Fig. 1.1 (**a**) Bloch sphere representing the coefficient of a qubit as in Eq. (1.2). A 3-dimensional vector pointing to a point on the sphere is called the spin variable $\boldsymbol{\sigma}$. (**b**) Representation of paths in quantum annealing. A single path on a 2-dimensional lattice (left) is represented by a 0/1 sequence on a 3-dimensional lattice

The coefficient $B_p = \left(A_p + G_p\right)$ for each term $|p\rangle$ interferes in amplitude, either increasing or decreasing based on the phase interference between the complex numbers A_p and G_p (Sect. 9.1.1). There are several known gate configurations $\hat{G}$ that increase the amplitude of the coefficient B_Q for the correct bit $p = Q$ while decreasing the amplitudes $\left|B_{p \neq Q}\right|$ for all other states [2, 4].[6] Observing the output $\hat{F} \cdot |\mathrm{Init}\rangle$ will then likely result in measuring the correct bit $|Q\rangle$ with high probability.

The coefficients of the qubit in Eq. (5.17) are generally parameterized as follows:

$$|\psi^{(2)}_{(\theta,\phi)}\rangle = \cos\frac{\theta}{2} \cdot |0\rangle + e^{i\phi} \sin\frac{\theta}{2} \cdot |1\rangle = \begin{pmatrix} \cos\frac{\theta}{2} \\ e^{i\phi} \sin\frac{\theta}{2} \end{pmatrix} \tag{1.2}$$

and are visualized as a 3-dimensional vector $\boldsymbol{\sigma}$ (spin variable) pointing to an azimuthal angle (θ, ϕ) on the Bloch sphere as shown in Fig. 1.1a (Sect. 7.2.1). The spin variable $\boldsymbol{\sigma}$ can be polarized and controlled in the same direction as an external magnetic field **B** pointing to the azimuthal angle (θ, ϕ) (Sect. 6.3.3). In this case, the qubit will have the coefficients as in Eq. (1.2) (Sect. 7.2.1).

Recalling that $c_0^{(1)} c_1^{(2)} c_0^{(3)} = A_{010}$ etc., the "change in each coefficients of Eq. (1.2) using magnetic field **B**" induces a change in A_p such that "$A_p \to G_p$". For an algorithm designed to extract the correct answer, $\hat{F} = \left(\hat{I} + \hat{G}\right)$" is correspondingly implemented so that the phase angle of complex coefficients G_p" is controlled using the magnetic field **B**, as described "quantum gate implementation". In practice, the azimuthal angle of Eq. (1.2) is often realized as the polarization

[6] You can learn about these typical examples as "quantum gate algorithms" from various texts (e.g., Grover's algorithm).

angle of photons, and the implementation of quantum gates is achieved through polarization angle control using beam splitters, as shown in recent developments [2].[7]

1.2.2 Quantum Annealing

Among the many combinatorial optimization problems, there are some that, fortunately, can be modeled such that 'the solution can be represented as a multi-dimensional array pattern of $|0\rangle$ and $|1\rangle$ (Sect. 8.2.1). For problems that can be modeled in this way, it is possible to apply the formulation of quantum annealing to solve the optimization problem. As an example, consider the problem of optimizing the 2D path taken by a moving object on a 2D grid [5] (Sect. 8.3.2). The position of the moving object at time t_k is represented by a bright spot $|1\rangle$, and other grid positions are represented by the value $|0\rangle$ [Fig. 1.1b]. By stacking these "0/1 patterns of the 2D grid" sheets for each time t_k, a 3-dimensional array of 0/1 patterns is created, which represents a single path. There are many possible paths, each of which is represented by a '3D block of 0/1 arrays'. Finding a single optimal path, therefore, maps to the problem of finding the optimal '0/1 array in the 3D block'.

Let the index of the grid position be $\mathbf{j}$. The value at each point can be expressed as follows, similar to Eq. (1.2):

$$|\psi^{(2)}\rangle_{\mathbf{j}} = |0\rangle_{\mathbf{j}} = \begin{pmatrix} 1 \\ 0 \end{pmatrix}, \quad |\psi^{(2)}\rangle_{\mathbf{j}} = |1\rangle_{\mathbf{j}} = \begin{pmatrix} 0 \\ 1 \end{pmatrix}$$

Then, one path can be identified with the 'pattern of spin configurations $\{\sigma_{\mathbf{j}}\}$ arranged on a 3D lattice' (Sect. 7.2.1). For certain problems, the objective function whose optimal spin configuration achieves the minimum value can be expressed in a quadratic form concerning the spins as[8] (Sect. 8.2.2):

$$\hat{H}_0 = -\sum_{i,j} J_{ij} \cdot \boldsymbol{\sigma}_i \cdot \boldsymbol{\sigma}_j \, . \tag{1.3}$$

In the Hamiltonian with an additional term to this,

$$\hat{H} = \hat{H}_0 - \boldsymbol{\Gamma} \sum_j \boldsymbol{\sigma}_j \, ,$$

[7] In addition to magnetic fields and polarization angles, various methods have been developed to realize and control qubits, including using the energy levels of ions and atoms [2].

[8] The vector notation for the index $\mathbf{j}$ has been omitted for simplicity.

this additional term acts to stabilize each $\boldsymbol{\sigma}_j$ in the direction of $\boldsymbol{\Gamma}$ due to the inner product. When we take $\boldsymbol{\Gamma}$ in the x-direction, the $\boldsymbol{\sigma}_j$ points to the x-direction, corresponding to $[(\theta, \phi) = (\pi/2, 0)]$ in Fig. 1.1a, which points sideways, hence $\boldsymbol{\Gamma}$ is called the transverse magnetic field .

According to Eq. (1.2), the spin $\boldsymbol{\sigma}_j$ pointing in the x-direction, corresponds to the quantum state,

$$|\psi^{(2)}_{\pi/2,0}\rangle = \begin{pmatrix} \cos\frac{\pi}{4} \\ e^{i0}\sin\frac{\pi}{4} \end{pmatrix} = \frac{1}{\sqrt{2}}(|0\rangle + |1\rangle) \ .$$

This corresponds to a state where each site is in a superposition of two possibilities, being either a bright spot or not. Therefore, minimizing the objective function $\hat{H}$ in the limit of a strong transverse magnetic field $\boldsymbol{\Gamma}$ results in a state that is a superposition of 'all possible paths'. As the transverse magnetic field is gradually weakened, the minimization solution of $\hat{H}_0$ is progressively selected from among all these possibilities.[9] In this way, quantum annealing achieves a global optimization solution by progressively narrowing down all possibilities without missing any. The essence of the difficulty in combinatorial optimization problems lies in finding the global optimal solution from among many local solutions. Quantum annealing overcomes this problem by applying and gradually weakening the transverse magnetic field.

As mentioned in this section with the phrase "if we are fortunate enough to model...", only a few among the many combinatorial optimization problems in the world can be handled by quantum annealing. This is possible only when the following mappings are established: "**(i)** the solution can be represented by a spin configuration $\{\sigma_j\}$", and "**(ii)** the objective function representing the optimal solution can be expressed in a quadratic form like Eq. (1.3)".[10] The number of problem groups for which such mappings have been successful is still quite limited [6, 7]. This is the most challenging aspect of promoting the practical use of quantum annealing.

1.3 Problem Setting for the Following Chapters

Provided an overview as above, the points that arise as unclear to beginners outside the field include:

---

[9] The theoretical framework that guarantees reaching the minimum solution in this manner is the theory of quantum annealing [6, 7].

[10] This is referred to as bringing it into the "QUBO (Quadratic Unconstrained Binary Optimization) form".

- **(a)** How did the superposition state like in Eq. (5.17) come about?
- **(b)** Why is the parameterization like in Eq. (1.2) used? Why is it $\theta/2$ instead of θ?
- **(c)** Why does the "state of $\theta/2$" in Eq. (1.2) polarize in the direction of an external magnetic field with angle θ?

Regarding **(a)**, there are tutorial methods that begin with experimental facts such as "double-slit interference" and "photon polarization beam splitters", thereby avoiding a detailed account of the long historical background to arrive at the idea of (a) [8]. Nowadays, these phenomena can be demonstrated as experimental facts,[11] so it is also possible to take the logical flow in such a way that "the mathematics framework to describe these quantum phenomena must be like this". In this book, to avoid redundancy with similar books, we have chosen to provide a minimum overview of how (a) came to be discussed, being traditional way.

As emphasized at the beginning of this chapter, I have never encountered a student who stopped at the level of the extent that leading physicists ponder when teaching the quantum mechanical superposition state (a). If the flow leading up to such "interpretational issues of quantum mechanics" is carefully explained, they could properly understand that the issue is surely making "leading physicists". Nowadays, with quantum computing as a definite motivation for learning, few students stop at the observation problem; understanding the difficulty of the observation problem, they move on. As detailed in Sect. 5.2.3, most students who are confused by the superposition state in Eq. (1.1) are actually confused at the level of hearing for the first time that "a general vector can be expanded in eigenvectors" due to a lack of understanding of elementary linear algebra.[12]

Regarding **(b)**, it is often brushed off with the terminology of group theory by saying "it's the correspondence between $SU(2)$ and $O(3)$" [9]. However, instead of dismissing it by saying "study a group theory textbook to understand this", this book attempts a more heartfelt and compact explanation. Regarding **(c)**, in materials science departments, it is often concluded with 'spin is like a magnetic moment, so...' and students get used to it after hearing it a few times. However, it is quite difficult to logically explain it when asked. It is unreasonable to impose this kind of habitual understanding on beginners in information science. However, trying to teach this leads to questions like 'why is spin a magnetic moment' and ultimately 'what is spin' itself. Again, instead of dismissing it by saying 'study a field theory textbook', I have tried structuring it in a way that even students from other fields can follow the logic by tracing back to the minimum necessary point.

---

[11] Until the 1980s, the double-slit experiment was a thought experiment proposed as a necessary consequence of quantum mechanics if its logic was correct.

[12] There were no students who claimed "I don't understand" at the level of wondering, "Is it possible to expand functions, which are infinite-dimensional vectors, in eigenvectors as we proved for finite-dimensional vectors?" which pertains to Hilbert space theory.

Due to limited space, I have focused on outlining the 'flow of concepts and ideas' that serve as a guidepost, and refer to detailed references. I aim to provide a 'bridge to further topics' that will make readers feel 'I want to learn more if I have time'.

1.4 Position and Structure of This Book

Readers, especially those not specialized in physics, may wonder why topics such as electromagnetism, relativity, and analytical mechanics are included in a book aimed at 'understanding quantum annealing'. It is necessary to explain the intention behind including these topics and how they are connected to the goal of the book.

1.4.1 Course for Understanding 'Why Such Topics Were Introduced'

In a society that values diversity, more attention should be paid to the diversity of learning methods, something the author has long believed. The author recalls that during my high school years, there was a widespread sentiment that 'teaching English grammar is bad', 'teaching formulas is bad', and a prevailing view that 'Japanese education has been wrong so far'. The phrase 'Just use and learn English, the reason Japanese people can't speak English is because they think about grammar', became pervasive. Education that emphasized 'just use/speak' English, like 'today's lesson is a travel scene', and teaching phrases for specific situations was strongly advocated.

Indeed, there are highly talented individuals who can master a foreign language fluently just by chatting at the kitchen table. For such people, traditional tutorials involving 'what is SVO structures' or 'what is present perfect tense' might have been objects of resentment. However, about half of the students seemed to learn by wondering, 'What is the role of this word?' and 'I can't understand the structure unless you label the roles as adjectives or objects'. For these students, being told, 'Just use and learn by your ear without worrying about the roles of words', would be confusing. Learning methods have diversity, so it is excessive to conclude that 'this way of teaching is bad'.

The contrast between the "formal scientific introduction to quantum mechanics" and the "traditional course to teach quantum mechanics" discussed in Sect. 1.1 feels structurally similar to the above story about English teaching methods. The traditional course was created with the intention of introducing 'how such topics were naturally developed through historical and contextual background'. However, I often see prefaces that seem to reject these traditional way to teach, saying, "This method is nothing but why it doesn't work". On the other hand, there are also books that call attention to the significance of 'courses that carefully introduce the background' even in the field of pure mathematics (typical formal science) [10, 11] and I feel that the "diversity of learning" is being ensured.

In understanding 'quantum annealing', which is the goal of this book, there are certainly some who find it more comfortable to follow tutorials that emphasize "tools that conform to such axioms exist as facts of nature, so let's focus on 'now' rather than 'past/background'". However, there is also a demand for learning methods that prefer to "understand the background story that can be likened to the 'past' before progressing". The inclusion of electromagnetism, relativity, and analytical mechanics in this book follows this intention.

1.4.2 Connection Between Chapters

Next, I will explain how the contents of each chapter are related to the ultimate goal of understanding quantum annealing, explaining in the way 'backwards from the goal'.

In the field of quantum computation, including quantum annealing, the key concept for beginners to grasp is the "superposition state." The appropriate mathematical expression for this is "spin." Since $|1\rangle$ represents an up-spin and $|0\rangle$ represents a down-spin, beginners initially tend to perceive spin as "something like polarization in three-dimensional space." However, they become confused when told that "a sideways spin is $\left[(|0\rangle + |1\rangle)/\sqrt{2}\right]$." To understand such mathematical concepts of **spin**, it is essential to grasp the mathematical concept of "field transformation under coordinate transformation".[13] Without a solid understanding of field transformation, elementary imagery will not suffice. Therefore, to ensure a proper understanding of field transformation, the chapter on mathematics (Chap. 2) begins with **linear algebra** (Fig. 1.2).

The main subject of this book involves elementary theoretical content in physics, with extensive use of mathematical derivation. The readers of this book are assumed to be non-specialists who have completed first-year university-level calculus and linear algebra but have not fully grasped the concepts. For those who struggle to follow the mathematical expressions used in physics and engineering and feel a sense of difficulty, the author identifies two main factors based on their teaching experience. One is the concept of **linearization** using Taylor expansion, and the other is the **index operations** such as matrix algorithms. In Chap. 2, these topics are carefully explained to ensure a thorough re-learning of linear algebra.

[13] Many teachers get used to these rules, but when asked by beginner students, "Why is that?" most teachers can only respond, "Read a book and study."

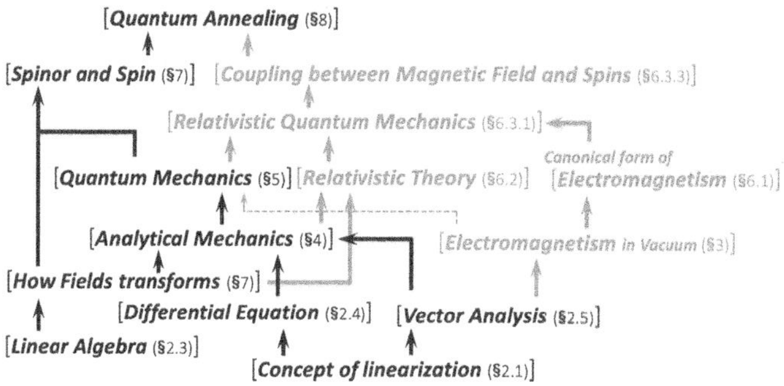

Fig. 1.2 The arrows indicate how each section of this book is related to the ultimate goal of understanding. The dotted arrows from "Electromagnetism in Vacuum" to "Quantum Mechanics" are necessary as background for discussing the optical-mechanical analogy, but the understanding of the equations in that context is not necessarily essential for later references

In quantum annealing and related areas, the control of quantum states is discussed, such as "orienting spin horizontally with a magnetic field." Here, those who are vaguely understanding that "spin is something like polarization in three-dimensional space?" will not question the idea that "spin aligns with the applied magnetic field." However, once they understand that "spin represents the coordinate transformation properties of a two-component field," they will question, "Why does such a transformation property align with the magnetic field direction? Where does this idea come from?" To logically connect this understanding, as shown in Fig. 1.2, it is necessary to briefly explain relativistic quantum mechanics. However, to explain this, a minimum explanation of **relativistic mechanics** and **electromagnetism** is required, otherwise, it will be impossible to avoid explanations like "According to so-and-so, it is known that..." and the book's purpose will not be achieved.

If you can proceed without asking "why?" regarding "spin polarizes in the same direction as the magnetic field", for the time being, the content indicated by the gray arrows in Fig. 1.2 can be detached from the course.[14] In this case, relativity and electromagnetism can be omitted from the course.

This book is, of course, written with the assumption that it will be read from the beginning, but it is also possible to read it in reverse, starting from the final Chap. 8 and working backwards. To ensure readers can trace back to where each topic is introduced, careful attention has been paid to guide them to the relevant sections and chapters previously appeared.

[14] In the author's lectures, this content is omitted depending on the progress and level of the students.

References

1. Singh C et al (2006) Improving students' understanding of quantum mechanics. Phys Today 59:43. https://doi.org/10.1063/1.2349732
2. Nielsen MA, Chuang IL (2013) Quantum computation and quantum information. Cambridge University Press, Cambridge. ISBN: 978-1107619197
3. Johnson G (2004) A shortcut through time: the path to the quantum computer. Knopf Doubleday Publishing Group, New York. ISBN: 978-0375726187
4. Mermin ND (2007) Quantum computer science: an introduction. Cambridge University Press, Cambridge. ISBN: 978-0521876582
5. Utimula K, Ichibha T, Prayogo G, Hongo K, Nakano K, Maezono R (2021) A quantum annealing approach to ionic diffusion in solids. Sci Rep 11:7261. https://doi.org/10.1038/s41598-021-86274-3
6. Das A, Chakrabarti BK (2005) Quantum annealing and related optimization methods. Springer, Heidelberg. ISBN: 978-3540279877
7. Tanaka S, Tamura R, Chakrabarti B.K (2017) Quantum spin glasses, annealing and computation. Cambridge University Press, Cambridge. ISBN: 978-1107113190
8. Susskind L, Friedman A (2015) Quantum mechanics: the theoretical minimum. Penguin, London. ISBN: 978-0141977812
9. Georgi H (1999) Lie algebras in particle physics: from isospin to unified theories. CRC Press, Boca Raton. ISBN: 978-0738202334
10. Needham T (2023) Visual complex analysis. Oxford University Press, Oxford. ISBN: 978-0192868923
11. Hairer E, Wanner G (2008) Analysis by its history. Springer, New York. ISBN: 978-0387770314

Chapter 2
Essentials of Basic Mathematical Tools

This chapter provides an overview of the minimum foundational mathematics needed to understand this book. Although it may seem lengthy, a thorough understanding here will make the later sections smoother to follow, so please take the time to go through it. Many students struggle to revisit the vast syllabus items from undergraduate mathematics when trying to relearn the basics necessary for fields like physics or engineering. From years of teaching students who found "formulas challenging" or felt "underprepared from undergraduate mathematics", the root issues tend to boil down to two key concepts: [(**A**) The idea of expanding and linearizing from an equilibrium point (Sect. 2.2)] and [(**B**) the handling of indices in mathematical expressions (Sect. 2.2)]. Failing to grasp these two points often leaves students viewing Taylor series expansions merely as "formulas to memorize before exams" and struggling to internalize them conceptually. Similarly, without mastering indices, the mathematics beyond linear projections (Sect. 2.3) remains difficult. In almost any applied context within the fields of physics or engineering, the mathematics encountered essentially boils down to just two concepts: the "concept of linearization" and the "mathematics of vector spaces". In a way, it's a repetitive pattern. Therefore, by thoroughly grasping just the two points mentioned above (**A** and **B**), much of the difficulty with foundational theories involving mathematical expressions in physics and engineering can be significantly alleviated.

Section 2.4 addresses differential equations and wave dynamics, which are recurring themes in engineering with a physics base. Many students continue into higher education with a shaky understanding of wave phenomena. From my experience teaching in information science departments, I encountered many students who would say, 'I understand waves in the time domain, but as soon as wave vector **k** appears, I ' m lost...'. If they could master the basic concept of mapping repetitive phenomena to a 'single rotation around a circle', they would find it manageable whether in time or space domains. However, since this concept is not taught thoroughly, students end up 'memorizing and using it without real understanding', which leads to hesitation when encountering unfamiliar expressions

R. Maezono, *Essentials for Deeper Understanding of Quantum Computing*,
https://doi.org/10.1007/978-981-96-5646-2_2

like $\exp(i\mathbf{k} \cdot \mathbf{r})$. The end of this chapter covers an explanation of vector analysis (Sect. 2.5). If one has thoroughly mastered the index operations in Sect. 2.2, they should be able to read through it without much difficulty.

The topics covered here are typically dealt with as subjects under 'mathematics'. 'Mathematics' is structured as a formal science with rigorous logical proofs and an appropriate formal style. However, the goal of this chapter is not to develop 'mathematics' itself but to explain it as a 'tool to be used' for discussions in later chapters. For this reason, I refer to it not as 'mathematics' but as 'basic mathematical tools'. Rather than orthodox and flat mathematics such as calculus and linear algebra, this chapter introduces a degree of narrative, at the expense of strict mathematical rigor.

2.1 The Concept of Linearization and Taylor Expansion

The concept of linearization represented by Taylor expansion form the foundation of mathematical physics and engineering mathematics. However, students who cannot readily recite and write out Taylor expansions are often observed. These students likely do not understand what the equations mean and, seeing the expansion formula as nothing more than a "chant", find it difficult to memorize.[1]

In essence, Taylor expansion is an implementation of the idea of "representing a general curve approximately as a power series polynomial in the vicinity of a point of interest". When learning existing theories in physics and engineering, simply grasping this concept is often sufficient. In this section, we'll take a relaxed approach to explain Taylor expansion based on this perspective. Throughout the explanation, we'll also add occasional remarks on "what might be considered insufficient from the viewpoint of pure mathematics".

2.1.1 From Linear Approximation to Taylor Expansion

No matter how curved a function may be, if it is smooth, we can assume that within a sufficiently small interval around a point $x = a$, it behaves like a straight line with the slope $f'(a)$. This is the intuition behind the concept of "differentiating a curve".

[1] Taylor expansion is often covered in "pure mathematics" within g eneral education courses. In these courses, the emphasis is placed more on acquiring the discipline of "developing logical arguments without ambiguity" rather than on understanding its applications. Professors teaching advanced courses in physics and engineering frequently assume students have "already learned this in general mathematics" and move on accordingly. This approach results in students who miss the opportunity to understand the meaning behind the equations, leaving them with an incomplete grasp of the concepts.

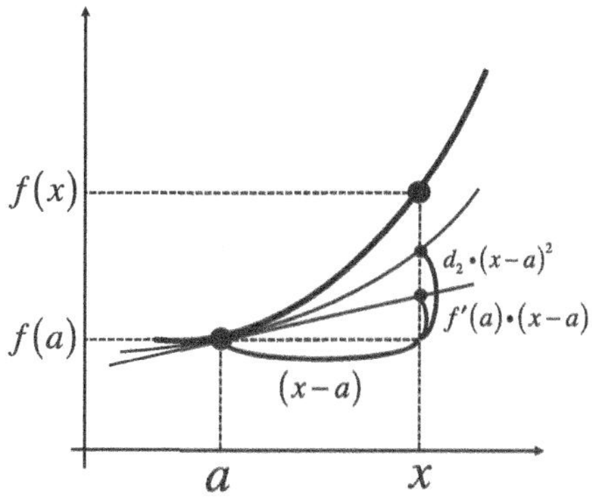

Fig. 2.1 To approximate $f(x)$, we start with a linear approximation. Next, we add a second-order term to refine the approximation. Then we proceed to third-order, fourth-order, and so on. By continuing this process, we can expect to achieve a more accurate approximation

This idea can be expressed in the formula, [2]

$$f(x) \approx f(a) + f'(a) \cdot (x - a) \ .$$

The "curvature" that cannot be captured by the above linear approximation can initially be represented by adding a second-order correction term:

$$f(x) \approx f(a) + f'(a) \cdot (x - a) + d_2 \cdot (x - a)^2 \ ,$$

as shown in Fig. 2.1.

The practical essence of Taylor expansion lies in expressing further errors with higher-order terms,[3] as follows:

$$f(x) \approx f(a) + f'(a) \cdot (x - a) + d_2 \cdot (x - a)^2 + d_3 \cdot (x - a)^3 + \cdots \ .$$

Following this idea, any function can be represented as a power series expansion:

$$f(x) = c_0 + c_1(x - a) + c_2(x - a)^2 + c_3(x - a)^3 + \cdots \ , \tag{2.1}$$

where we set up the problem to determine each coefficient $\{c_j\}$.[4]

---

[2] Note that rearranging this formula yields the "definition of a derivative" as learned in high school.

[3] A quadratic function has only one inflection point, so the correction will continue to increase monotonically. To express a correction that "first increases and then decreases", a cubic function with two inflection points is necessary. Furthermore, to express a correction that "increases, then decreases, and then increases again", a quartic function is required. This logic of adding higher-order terms reflects the spirit behind such an approach.

[4] $\{c_j\}$ represents the set $\{c_1, c_2, \cdots\}$. It is essential to develop the habit of properly using such notation.

First, by substituting $x = a$ into Eq. (2.1) and comparing both sides, we find that $c_0 = f(a)$. Next, taking the first derivative of both sides yields:

$$f'(x) = c_1 + 2 \cdot c_2(x-a) + 3 \cdot c_3(x-a)^2 + \cdots .$$

Substituting $x = a$ and comparing both sides gives $c_1 = f'(a)$. Taking further the second derivative of both sides and similarly substituting $x = a$ for comparison yields $2 \cdot c_2 = f''(a)$.

By applying the similar procedure of "taking the k-th derivative, substituting $x = a$, and comparing both sides", each coefficient $\{c_j\}$ can be extracted as follows:

$$c_0 = f(a)\ ,\ c_1 = f'(a)\ ,\ c_2 = \frac{1}{2} f''(a)\ ,\ c_3 = \frac{1}{2 \cdot 3} f^{(3)}(a)\ ,\ \cdots .$$

Substituting these coefficients back into Eq. (2.1) gives:

$$\begin{aligned} f(x) = f(a) + f'(a)(x-a) + \frac{1}{2} f''(a)(x-a)^2 + \frac{1}{2 \cdot 3} f^{(3)}(a)(x-a)^3 + \cdots \\ \cdots + \frac{1}{k!} f^{(k)}(a)(x-a)^k + \cdots . \end{aligned} \tag{2.2}$$

This is the well-known **Taylor expansion**. It is essential to write this formula out several times until you can **recite and reproduce it from memory**.

Taking Notes by Hand

If your goal is to truly overcome a "math formula aversion", it is strongly recommended that you copy the equations in this chapter into a notebook, writing them out by hand. In the past, the value of such handwritten note-taking was questioned under the idea of "is there any meaning to this kind of rote transcription?" and lecture styles featuring "PowerPoint presentations and handouts" (printed lecture materials handed to students) became the norm. However, this shift seems to have coincided with an increase in students who struggle with mathematical formulas. Having taught the content of this chapter for many years, I made it a rule to prohibit the use of handouts or mobile devices during lectures, encouraging students to focus on writing things down by hand, even if it felt like mere transcription. As a result, many students naturally became proficient in handling formulas. Just as "you cannot memorize a word you cannot pronounce", there seems to be a similar principle: "you cannot become proficient with a formula you cannot write neatly yourself.

In the case where $a = 0$ in Eq. (2.2), the expression

$$f(x) = f(0) + f'(0) \cdot x + \frac{1}{2} f''(0) \cdot x^2 + \frac{1}{2 \cdot 3} f'''(0) \cdot x^3 + \cdots$$
$$\cdots + \frac{1}{k!} f^{(k)}(0)\, x^k + \cdots, \tag{2.3}$$

is called the **Maclaurin series**. This series represents a polynomial approximation of the "behavior around the origin" for dependencies expressed by a curve.

The "deviation from the point of interest $x = a$" can be expressed as $\delta x = x - a$. Thus, since $x = a + \delta x$, the Taylor expansion, Eq. (2.2), can be rewritten as:

$$f(a + \delta x) = f(a) + f'(a) \cdot \delta x + \frac{1}{2} f''(a) \cdot \delta x^2 + \frac{1}{2 \cdot 3} f^{(3)}(a) \cdot \delta x^3 + \cdots$$
$$\cdots + \frac{1}{k!} f^{(k)}(a) \cdot \delta x^k + \cdots. \tag{2.4}$$

2.1.2 *Truncating an Expansion as a Standard Approximation*

The practical utility of Taylor expansion lies in what is known as **truncating the expansion**, as described below. For example, consider the processing of voltage signals by an amplifier. If we imagine handling a "small signal δx" around a constant voltage $x = a$, the smallness of the signal allows us to assume $\delta x \ll 1$.[5] Consequently, we can assert $\delta x \gg \delta x^2 \gg \delta x^3 \gg \cdots$, enabling us to **ignore higher-order contributions**. With this perspective in mind, the Taylor expansion, Eq. (2.4) is often expressed as:

$$f(a + \delta x) = f(a) + f'(a) \cdot \delta x + \frac{1}{2} f''(a) \cdot \delta x^2 + O\left(\delta x^3\right). \tag{2.5}$$

Here, $O\left(\delta x^3\right)$ indicates "terms with higher order than the third δx", with the implicit intention of saying, "these contributions are small enough to neglect". This approach allows us to quantitatively grasp the behavior of the system up to a quadratic approximation. The behavior of quadratic functions is thoroughly studied in middle and high school, and by applying that knowledge, we can analyze the behavior of systems with respect to small variations.[6] Truncating the expansion at the quadratic

[5] If we define the characteristic scale (e.g., length) of the focus x as L, scaling x as $\xi = x/L$ (normalizing its magnitude according to the scale) allows us to describe the quantitative variation using ξ. In this context, $\xi \sim 1$ represents "ordinary" magnitude, while $\xi \ll 1$ represents "small" magnitude, which can simplify the problem.

[6] This serves as an answer to middle and high school students who question why they are made to study quadratic functions and struggle to find motivation in learning mathematics.

term is often referred to as **harmonic approximation** or **Gaussian approximation**, depending on the context. The latter term is particularly used when truncating a function exponentiated within a Gaussian distribution to the quadratic term.

The most commonly used approximation is truncating the expansion at the first-order term:

$$f(a+\delta x) = f(a) + f'(a) \cdot \delta x + O\left(\delta x^2\right) . \tag{2.6}$$

This approximation, which represents a linear function (a straight line), is referred to as **linear approximation**. In this case, the response output to a small input can be treated as proportional to the input, significantly simplifying the analysis. While simplifying the behavior of individual responses, this approximation enables the analysis of large-scale system chains as a composite of these simplified individual responses.

The approach described above, namely, approximating behavior using linear or quadratic functions around a point of interest, is a standard method that repeatedly appears in physics and engineering. For small signal variations around the point of interest, as long as the system does not exhibit problematic responses such as being non-differentiable or having discontinuities, it is meaningful to linearize the response using Taylor expansion. In such cases, the theoretical description of the response reduces to the same mathematical form, regardless of the subject being described. This type of **linear theory** serves as the foundation for mathematical theories in any field.[7] In linear systems, it can generally be stated that the response to "superimposed inputs" can be written as the "superposition of individual responses."[8]

What Do Mathematicians Worry About?

As explained above, we derived Taylor expansion without overly complex logical steps. However, in typical undergraduate education, the explanation is often more intricate-enough so that many students struggle and fall behind [1]. It is quite beneficial at this stage revisiting why such complexity is introduced. Taylor expansion takes the form of an infinite series, so it is necessary to verify whether it genuinely provides "meaningful numerical values". On the other hand, this book explains the topic without such careful examination, proceeding with the approach like "by rewriting the equation,

(continued)

[7] Linear theory is used as the fundamental framework, and modifications are added to expand the theory to handle more general situations.

[8] Such linear theory has been regarded as a subject of general education, serving as foundational knowledge that individuals receiving higher education in the sciences and engineering, regardless of their specific field, are expected to acquire.

this formula is derived...". This kind of argument is sometimes referred to as **just a formal** argument. To ensure a rigorous discussion, one must carefully evaluate the remainder term when truncating the series and demonstrate that the remainder approaches zero in the limit. Such evaluations typically rely on the **Intermediate Value Theorem** or the **Mean Value Theorem** . Proving these theorems requires precise mathematical definitions of 'limits' and 'neighborhoods'. When considering the relationship $x \rightarrow f(x)$ and introducing concepts like "limits and neighborhoods" for the "projection" $f(x)$, the question arises: how do we ensure the consistent existence of $f(x) \rightarrow x$ as the origin of the projection? Addressing this in its entirety demands an extended and intricate discussion, which can lead to many students falling behind. To avoid such dropouts, one efficient approach is to first focus on reaching the "ridge trail" where students feel, "I understand the big picture". Once that sense of success is achieved, they can revisit and critically examine, "Was that really sufficient"? This book adopts the style of avoiding detailed discussions and progressing formally to reach the "ridge trail" as quickly as possible. However, it is important to understand that the omitted "detailed discussions" are not mere minor details but rather the **most fundamental aspects.**[a]

[a]For example, verifying whether the series provides meaningful values is fundamental.

2.1.3 *Euler's Formula*

By applying the Maclaurin series in Eq. (2.3) to trigonometric and exponential functions, we obtain the following expansions:

$$\sin x = x - \frac{1}{3!}x^3 + \frac{1}{5!}x^5 - \cdots ,$$

$$\cos x = 1 - \frac{1}{2!}x^2 + \frac{1}{4!}x^4 - \cdots ,$$

$$\exp[x] = 1 + x + \frac{1}{2!}x^2 + \frac{1}{3!}x^3 + \frac{1}{4!}x^4 + \cdots .$$

By 'formally' substituting ix into the expansion of the exponential function, we have:

$$\begin{aligned}\exp[ix] &= 1 + ix - \frac{1}{2!}x^2 - i\frac{1}{3!}x^3 + \frac{1}{4!}x^4 + i\frac{1}{5!}x^5 \cdots \\ &= \left(1 - \frac{1}{2!}x^2 + \frac{1}{4!}x^4 + \cdots\right) + i\left(x - \frac{1}{3!}x^3 + \frac{1}{5!}x^5 + \cdots\right) \\ &= \cos x + i \cdot \sin x \end{aligned} \tag{2.7}$$

This is known as the **Euler's formula,**[9] a remarkably useful and elegant result.

Formal Derivation

It is crucial to develop an awareness that the "derivation" presented here is merely "formal". The "formal derivation" of the Taylor expansion in the previous section made sense in the context of real numbers, where it corresponded to an intuitive understanding of "distances on a diagram". As such, one must pause and ask, "Who permitted the application of such 'derivations' to imaginary numbers, $x' = ix$"? This reflective attitude is essential.[b] Although formal explanations may suffice for the purpose of "quickly tracing well-established theories already known to be correct", this approach can fail spectacularly when one attempts to build a new mathematical framework from scratch. You may encounter situations where each step in a calculation appears "correct" when lifted from formula collections, but the result leads to paradoxes. The opposite of "formal derivation" is "careful derivation", a practice that undoubtedly produces its share of dropouts every year. Yet, one must not resent this rigor. This discipline represents the very distinction between "technical training aimed at mastering existing frameworks" and "higher education that universities should provide to cultivate individuals capable of creating entirely new framework.

[b] When contemplating the meaning of exponential functions extended to complex numbers, one can find numerous insightful perspectives. It becomes clear why formal derivations might be considered an affront to Euler's genius [2].

2.2 Getting Familiar with Index Handling

Unfortunately, in recent times, it is not uncommon to see students mistakenly calculating the "dot product of two vectors" as a vector or treating the "product of a vector and a matrix" as a matrix. This reflects how many students proceed

[9] Leonhard Euler/1707.4–1783.9.

with an inadequate understanding of vectors and matrices. The three-dimensional vectors taught in high school are quantities that combine "three components into a single entity". These components are typically expressed using indices, such as $\{a_1, a_2, a_3, \cdots\} = \{a_j\}$ where $j = 1, 2, \cdots$. One effective way to approach vectors and matrices is to become proficient in "breaking them down into components a_j and explicitly working with them."[10] In this context, operations involving unfamiliar symbols like ε_{ijk} will appear. While these rules may seem intimidating at first, once you focus on mastering them, subsequent component-based calculations simplify into standard real-number operations. This eliminates the confusion caused by the variety of specialized rules in vector calculus. In the chapters that follow, the treatment of vectors and matrices will appear repeatedly. The content discussed in this section is foundational for understanding the subsequent material. Conversely, by carefully understanding this section, you will find that the projections and operations involving equations in later chapters become repetitive and relatively easy to handle.

2.2.1 Indices for Vectors and Matrices

In high school mathematics, a 3-dimensional vector can be written as:

$$\mathbf{a} = \begin{pmatrix} a_x \\ a_y \\ a_z \end{pmatrix} = a_x \cdot \mathbf{e}_x + a_y \cdot \mathbf{e}_y + a_z \cdot \mathbf{e}_z \ ,$$

where $\{\mathbf{e}_x, \mathbf{e}_y, \mathbf{e}_z\}$ are referred to as the basis vectors. Hereafter, we adopt the convention of replacing the component indices (x, y, z) with $(1, 2, 3)$. By doing so, we extend this concept to an N-dimensional space where the number of basis vectors is increased to N.

By adopting the convention of replacing component indices $(x, y, z, \cdots)$ with $(1, 2, 3, \cdots)$, a vector can be expressed as follows:

$$\begin{aligned} \mathbf{a} &= \begin{pmatrix} a_1 \\ a_2 \\ \vdots \\ a_N \end{pmatrix} = a_1 \cdot \mathbf{e}_1 + a_2 \cdot \mathbf{e}_2 + \cdots \\ &= \sum_k a_k \cdot \mathbf{e}_k \ , \end{aligned} \tag{2.8}$$

[10] This approach is sometimes referred to as tensor analysis [3]. However, for beginners, opening a textbook on "tensor analysis" may expose them to abstract and complex discussions that appear entirely unrelated to their expectations. While the term "tensor" will appear later in this text, introducing it too early risks confusing beginners. Thus, such terminology is avoided at this stage.

represented compactly using the summation symbol. The components are expressed with indices as $\{a_1, a_2, a_3, \cdots\} = \{a_j\}$, where $j = 1, 2, \cdots$.[11] One effective approach to mastering vectors and matrices is to "become proficient in writing expressions with awareness of the components a_j". Hereafter for a vector,

$$\mathbf{a} \sim \{a_j\} ,$$

we emphasize the importance of developing a habit to associate "the quantity a_j with index j" with a vector. Similarly, for a matrix,

$$A = \begin{pmatrix} a_{11} & a_{12} & \cdots & a_{1N} \\ a_{21} & & \ddots & \vdots \\ \vdots & & & \\ a_{N1} & \cdots & & a_{NN} \end{pmatrix} ,$$

we adopt the shorthand:

$$A \sim \{a_{ij}\} ,$$

and emphasize the habit of associating "the quantity a_{ij} with indices i and j" with a matrix.

The operation of taking the dot product,

$$\mathbf{A} \cdot \mathbf{B} = A_1 B_1 + A_2 B_2 + A_3 B_3 + \cdots = \sum_{l=1}^{N} A_l B_l ,$$

involves the process of "summing over the matching index l from 1 to N."[12] This can be compactly written as $\mathbf{A} \cdot \mathbf{B} = A_l B_l$, where we adopt the **Einstein summation convention** : whenever a matching index appears, it implies summation over that index.[13] Using this convention, Eq. (2.8) can be rewritten as:

$$\mathbf{a} = a_l \cdot \mathbf{e}_l .$$

---

[11] Becoming accustomed to using "set notation with curly braces" in this manner is an essential first step in mastering the theoretical courses.

[12] The concept of the vector inner product is likely introduced rather abruptly in high school courses. This is a good opportunity to reconsider "the natural context in which it is introduced", as summarized in Sect. 9.2.1. Additionally, it may not seem obvious why the inner product, defined as the "sum of component-wise products", results in the relationship: $\mathbf{a} \cdot \mathbf{b} = |\mathbf{a}| \cdot |\mathbf{b}| \cdot \cos\theta$, where θ is the angle between the vectors. For a detailed explanation of this, refer to the Chap. 9.

[13] Albert Einstein (1879.03–1955.04).

It is important to become accustomed to recognizing that the right-hand side of this expression implies the summation in Eq. (2.8).

In the context of Einstein's summation convention, the variable l, which "runs" from 1 to N, is referred to as a **dummy index**. Remembering a definite integral,

$$I = \int_a^b g(t) \cdot dt = \int_a^b g(y) \cdot dy = \int_a^b g(x) \cdot dx ,$$

the choice of the integration variable, whether t, x, or y, is arbitrary. These variables are also referred to as **dummy variables**. Similarly, for the dot product,

$$\mathbf{A} \cdot \mathbf{B} = A_l B_l = A_m B_m , \tag{2.9}$$

the choice of symbol for the dummy variable (such as l or m) is entirely arbitrary.[14]

Considering the expressions,

$$I = \sum_{l=1}^{3} A_l B_l = A_l B_l \quad , \quad J = \sum_{l=1}^{3} C_l D_l = C_l D_l],$$

it is crucial to avoid making the mistake of transcribing them directly as

$$I \cdot J = A_l B_l \cdot C_l D_l \quad \text{(Wrong !)} . \tag{2.10}$$

The correct interpretation is

$$I \cdot J = (A_1 B_1 + \cdots + A_3 B_3) \cdot (C_1 D_1 + \cdots C_3 D_3) ,$$

which is distinct from the misinterpreted result in Eq. (2.10):

$$A_l B_l \cdot C_l D_l = A_1 B_1 C_1 D_1 + A_2 B_2 C_2 D_2 + A_3 B_3 C_3 D_3 .$$

The key point is to be mindful of **which dummy variables pair together**, and to note that **dummy variables are unrelated to one another**. To ensure clarity, distinct symbols should be used for each dummy variable. Thus, the correct operation for Eq. (2.10) is:

$$I \cdot J = A_l B_l \cdot C_s D_s .$$

Here are some common mistakes that can occur due to unfamiliarity with Einstein's summation convention. Consider the spatial coordinates (x, y, z), written as

[14] In this book, we generally use l, m, or n for such purposes.

(r_1, r_2, r_3) by convention. Partial derivatives such as $\partial/\partial x, \partial/\partial y, \cdots$ are expressed as $\partial_j = \partial/\partial r_j$. In this context, one might mistakenly write:

$$\partial_l \left(r_l\right) = \frac{\partial}{\partial r_l}\left(r_l\right) = 1 \quad \text{(Wrong !)} \ .$$

However, recognizing that the index l implies a summation over it, the correct expression is:

$$\partial_l \left(r_l\right) = \sum_{l=1}^{N} \frac{\partial}{\partial r_l}\left(r_l\right) = \sum_{l=1}^{N} 1 = N \ . \tag{2.11}$$

Thus, the correct result is $\partial_l(r_l) = N$.

2.2.2 Diagonal Matrix and Kronecker Delta Notation

The **Kronecker Delta** notation[15] is defined by the convention:

$$\delta_{ij} = \begin{cases} 1 & ; \quad (i = j) \\ 0 & ; \quad (i \neq j) \end{cases} \ .$$

Since this corresponds to

$$\begin{pmatrix} \delta_{11} & \delta_{12} & \delta_{13} \\ \delta_{21} & \delta_{22} & \delta_{23} \\ \delta_{31} & \delta_{32} & \delta_{33} \end{pmatrix} = \begin{pmatrix} 1 & 0 & 0 \\ 0 & 1 & 0 \\ 0 & 0 & 1 \end{pmatrix} ,$$

it represents the "matrix elements of the identity matrix".

For the basis vectors appearing in Eq. (2.8),

$$\mathbf{e}_1 \cdot \mathbf{e}_2 = \begin{pmatrix} 1 \\ 0 \\ 0 \end{pmatrix} \begin{pmatrix} 0 \\ 1 \\ 0 \end{pmatrix} = 0 \quad , \quad \mathbf{e}_1 \cdot \mathbf{e}_1 = 1 \quad etc.,$$

the "orthogonality of the basis" can be expressed using the Kronecker delta as,

$$\mathbf{e}_i \cdot \mathbf{e}_j = \delta_{ij} \ . \tag{2.12}$$

[15] Leopold Kronecker (1823.12–1891.12).

It is essential to **familiarise oneself operationally** with the following relation,

$$\delta_{i\underline{l}} \cdot D_{st\underline{l}m} = D_{st\underline{i}m} \ . \tag{2.13}$$

By explicitly writing out the summation over the dummy variable, this equation becomes,

$$\begin{aligned}\sum_{l} \delta_{il} \cdot D_{stlm} &= \delta_{i1} \cdot D_{st1m} + \delta_{i2} \cdot D_{st2m} + \delta_{i3} \cdot D_{st3m} \\ &= \begin{cases} D_{st1m} & \cdots \quad (i = 1) \\ D_{st2m} & \cdots \quad (i = 2) \\ D_{st3m} & \cdots \quad (i = 3) \end{cases} \\ &= D_{stim} \ ,\end{aligned}$$

being clear in its **meaning**. Then, by revisiting Eq. (2.13), you should train yourself to **intuitively** understand this **as a manipulation** in the following way: While i is a fixed index, l functions as a dummy variable, "running" through $l = 1, 2, 3$. The idea is that, during this process, δ_{il} "pins down" l only where $l = i$ (fixed), thereby picking up only the term for $l = i$. Practice becoming proficient at transforming the left-hand side into the right-hand side with this concept in mind.

Think of δ_{il} as an operation that pins the floating index l to the fixed index i. Develop the ability to interpret this symbol with that intuition in mind. Moving forward, it is crucial to always be aware of "which variable is the dummy and which is the given or fixed subscript".

2.2.3 *Cross Product and Eddington's Epsilon*

These days, it is increasingly common to encounter students who are not well-versed in the concept of the vector cross product. As a review, the definition of the cross product is

$$\mathbf{A} \times \mathbf{B} = \begin{vmatrix} \mathbf{e}_1 & \mathbf{e}_2 & \mathbf{e}_3 \\ A_1 & A_2 & A_3 \\ B_1 & B_2 & B_3 \end{vmatrix} \tag{2.14}$$

$$= \mathbf{e}_1 (A_2 B_3 - A_3 B_2) + \mathbf{e}_2 (A_3 B_1 - A_1 B_3) + \mathbf{e}_3 (A_1 B_2 - A_2 B_1)$$

$$= \begin{pmatrix} A_2 B_3 - A_3 B_2 \\ A_3 B_1 - A_1 B_3 \\ A_1 B_2 - A_2 B_1 \end{pmatrix} .$$

It is important to commit the first expression to memory thoroughly.

From the second line of the above equation, we can express it in the form of a summation over l as

$$\mathbf{A}\times\mathbf{B} = \mathbf{e}_1\left(A_2B_3 - A_3B_2\right) + \mathbf{e}_2\left(A_3B_1 - A_1B_3\right) + \mathbf{e}_3\left(A_1B_2 - A_2B_1\right)$$
$$\sim \sum_{l=1}^{3} \mathbf{e}_l\left(A_sB_t - A_tB_s\right) . \tag{2.15}$$

For the temporarily assigned index triplets (l, s, t),

$$(l, s, t) = (1, 2, 3), (2, 3, 1), (3, 1, 2) ,$$

representing a relationship where the "cycle" rotates clockwise around the vertex (123), called **cyclic**. If this "cycle" instead rotates counterclockwise, the assignments become

$$(l, s, t) = (1, 3, 2), (2, 1, 3), (3, 2, 1) .$$

Such an assignment is referred to as **anti-cyclic**.

Now, let us take a closer look at the first line of Eq. (2.15). Considering the general term of the appearing terms as $\mathbf{e}_l \cdot A_sB_t$, it can be observed that when the index set (l, s, t) is cyclic, a positive sign is applied, whereas a negative sign is applied for an anti-cyclic index set. By introducing the symbol (known as **Eddington's symbol,**[16])

$$\varepsilon_{lst} = \begin{cases} +1 \; ; & (lst) \text{ is cyclic} \\ -1 \; ; & (lst) \text{ is anti-cyclic} \\ 0 \; ; & \text{otherwise} \end{cases} , \tag{2.16}$$

we can express the cross product as

$$\mathbf{A}\times\mathbf{B} = \sum_{l=1}^{3}\mathbf{e}_l\cdot\sum_{s=1}^{3}\sum_{t=1}^{3}\varepsilon_{lst}A_sB_t = \sum_{l,s,t}\mathbf{e}_l\cdot\varepsilon_{lst}A_sB_t . \tag{2.17}$$

For practice, when explicitly writing out and verifying, it leads to[17]

$$\mathbf{A}\times\mathbf{B} = \sum_{l=1}^{3}\mathbf{e}_l\cdot\sum_{s=1}^{3}\sum_{t=1}^{3}\varepsilon_{lst}A_sB_t$$
$$= \mathbf{e}_1\cdot\sum_{s=1}^{3}\sum_{t=1}^{3}\varepsilon_{1\cdot st}A_sB_t + \mathbf{e}_2\cdot\sum_{s=1}^{3}\sum_{t=1}^{3}\varepsilon_{2\cdot st}A_sB_t$$

[16] Arthur Stanley Eddington (1882.12–1944.11).

[17] In the final line, only the summation over $\mathbf{e}_1$ is expanded, while those over $\mathbf{e}_2$ and $\mathbf{e}_3$ are left unexpanded. Einstein's summation convention is applied, omitting explicit summation symbols.

$$
+\mathbf{e}_3 \cdot \sum_{s=1}^{3}\sum_{t=1}^{3} \varepsilon_{3\cdot st} A_s B_t
$$

$$
= \mathbf{e}_1 \cdot \begin{pmatrix} \varepsilon_{1\cdot 11} A_1 B_1 + \varepsilon_{1\cdot 12} A_1 B_2 + \varepsilon_{1\cdot 13} A_1 B_3 \\ +\varepsilon_{1\cdot 21} A_2 B_1 + \varepsilon_{1\cdot 22} A_2 B_2 + \varepsilon_{1\cdot 23} A_2 B_3 \\ +\varepsilon_{1\cdot 31} A_3 B_1 + \varepsilon_{1\cdot 32} A_3 B_2 + \varepsilon_{1\cdot 33} A_3 B_3 \end{pmatrix}
$$

$$
+\mathbf{e}_2 \cdot \varepsilon_{2st} A_s B_t + \mathbf{e}_3 \cdot \varepsilon_{3st} A_s B_t \, . \tag{2.18}
$$

Applying Eddington's rule, Eq. (2.16), it is

$$
= \mathbf{e}_1 \cdot \begin{pmatrix} 0 \cdot A_1 B_1 + 0 \cdot A_1 B_2 + 0 \cdot A_1 B_3 \\ +0 \cdot A_2 B_1 + 0 \cdot A_2 B_2 + (+1) \cdot A_2 B_3 \\ +0 \cdot A_3 B_1 + (-1) \cdot A_3 B_2 + 0 \cdot A_3 B_3 \end{pmatrix}
$$

$$
+\mathbf{e}_2 \cdot \varepsilon_{2st} A_s B_t + \mathbf{e}_3 \cdot \varepsilon_{3st} A_s B_t
$$

$$
= \mathbf{e}_1 \cdot (A_2 B_3 - A_3 B_2) + \mathbf{e}_2 \cdot \varepsilon_{2st} A_s B_t + \mathbf{e}_3 \cdot \varepsilon_{3st} A_s B_t \, .
$$

The first term involving $\mathbf{e}_1$ indeed matches the corresponding term in Eq. (2.15). Similarly, for the second and third terms, the property of Eddington's rule (where only cyclic or anti-cyclic terms contribute, and all others vanish) ensures that the summation over s and t in Eq. (2.17) holds validly being coinciding with Eq. (2.15).

By applying Einstein's summation convention to the summation over s and t in Eq. (2.17), we can write

$$
\mathbf{A} \times \mathbf{B} = \sum_{l=1}^{3} \mathbf{e}_l \cdot \varepsilon_{lst} A_s B_t
$$

$$
= \mathbf{e}_1 \cdot \varepsilon_{1st} A_s B_t + \mathbf{e}_2 \cdot \varepsilon_{2st} A_s B_t + \mathbf{e}_3 \cdot \varepsilon_{3st} A_s B_t . \, .
$$

The "j-th component of the cross product" can hence be expressed as

$$
[\mathbf{A} \times \mathbf{B}]_j = \varepsilon_{jst} A_s B_t \, . \tag{2.19}
$$

Regarding Eq. (2.19), once you thoroughly understand the semantic validity of the equality, it is important to internalize the expression as a procedural tool. Practice writing it out repeatedly until you can handle it mechanically.[18] In particular, focus on understanding the roles of each index in the equation. For a fixed j, s and t serve

[18] It is crucial to manually write out the multi-level summation over (l, s, t) in the rightmost expression of Eq. (2.17) at least once, as shown in Eq. (2.18). This practice will help you visualize such multi-level summation expressions and link them to their explicit expansions like Eq. (2.18). Some students fail to interpret expressions like $\sum_{ij} a_{ij}$ and see them as meaningless symbols rather than mathematical constructs.

as dummy variables under the summation. Within the summation, only the terms where s and t are cyclic or anti-cyclic with respect to the given j contribute with positive or negative signs. Commit this mechanical relationship between indices to memory. This equation will appear repeatedly in future contexts, so it is important to be comfortable with it.

To solidify your understanding of Eddington's symbol and Einstein summation convention, let us confirm that the determinant of a 3×3 matrix can be expressed as

$$\begin{vmatrix} a_{11} & a_{12} & a_{13} \\ a_{21} & a_{22} & a_{23} \\ a_{31} & a_{32} & a_{33} \end{vmatrix} = \varepsilon_{ijk} \cdot a_{1i}a_{2j}a_{3k} \,. \tag{2.20}$$

Expanding the summation using Einstein's convention, the right-hand side includes all possible combinations of (i, j, k), resulting in $3 \times 3 \times 3 = 27$ terms. Among these, the terms where Eddington's symbol is non-zero occur only when all indices (i, j, k) are distinct. Consequently, there are only six terms in total: three "cyclic permutations" with a positive sign and three "anti-cyclic permutations" with a negative sign. By assigning specific values of (i, j, k) to $a_{1i}a_{2j}a_{3k}$ and applying the corresponding positive or negative signs, we arrive at the well-known formula for the determinant of a 3×3 matrix.

2.2.4 *Processing Products of Eddington's Symbols*

One of the most useful formulas for practical applications is the following. It is crucial to thoroughly understand the explanation below and practice writing it out several times to memorize it well:

$$\varepsilon_{jst}\varepsilon_{jmn} = \delta_{sm}\delta_{tn} - \delta_{sn}\delta_{tm} \,. \tag{2.21}$$

To straightforwardly demonstrate this formula, we consider the summation over the index j:

$$\varepsilon_{jst}\varepsilon_{jmn} = \varepsilon_{1st}\varepsilon_{1mn} + \varepsilon_{2st}\varepsilon_{2mn} + \varepsilon_{3st}\varepsilon_{3mn} \cdot . \tag{2.22}$$

By explicitly writing out all 3^4 combinations of (s, t, m, n) in a table and comparing their values with the right-hand side of Eq. (2.21), we can verify the equivalence. However, even without resorting to such a brute-force approach, it is immediately apparent that the three terms on the right-hand side of Eq. (2.22) are **mutually exclusive** in the sense that "if one term is 1, the other two must be zero". For instance, the first term becomes non-zero only for the pair $(s, t) = (2, 3)$, and this ensures that the second and third terms are zero. Only one term contributes a value, and it is always either 0 or ± 1, which are the only possible outcomes for the product of ε_{ijk} symbols. For example, let us consider the first term of Eq. (2.22), $\varepsilon_{1st}\varepsilon_{1mn}$. When it is non-zero, such as for the pair $\{s, t\} = \{2, 3\}$, specific cases like $\varepsilon_{123}\varepsilon_{123}$, $\varepsilon_{132}\varepsilon_{132}$, or $\varepsilon_{123}\varepsilon_{132}$ arise. The product takes a positive value when both ε_{ijk} symbols have the same sign, i.e., "positive $\times$ positive" or "negative $\times$ negative". This occurs when the two Eddington symbols are either both cyclic or both anti-cyclic, corresponding to $s = m$ and $t = n$. These cases are represented by the term $\delta_{sm}\delta_{tn}$, contributing positively. Similarly, the left-hand side takes a negative value when represented by the term $\delta_{sn}\delta_{tm}$, which occurs when one symbol is cyclic and the other is anti-cyclic. Recognizing that these two cases are mutually exclusive, we conclude that the values on the right-hand side of Eq. (2.22) can be equivalently represented by the delta symbols on the right-hand side of Eq. (2.21).

Equation (2.21), like Eqs. (2.13) and (2.19), should first be thoroughly understood semantically to ensure the equality holds. Once this is done, practice writing it out several times to internalize it, enabling mechanical application as an operational tool. Formulas are much easier to memorize when their "interpretation" is clear.[19] The **reading/memorization** of Eq. (2.21) is as follows: First, look at the subscripts of the two ε symbols on the left-hand side and ensure that, in both, the subscript j is positioned at the beginning. Then, focus on the subscripts other than j and recite, "If $st = mn$, the two ε symbols have the same sign", or "If $st = nm$, the two ε symbols have opposite signs". By repeatedly practising writing out the Dirac's delta on the right-hand side while chanting this, you can memorize it. This equation will also frequently appear in later discussions.

If, on the left-hand side of Eq. (2.21), the subscripts of the two ε symbols are not positioned with j at the beginning in both cases, then, for instance:

$$\varepsilon_{jst}\varepsilon_{mnj} = \varepsilon_{jst}\varepsilon_{jmn}$$

cyclically permute the subscripts of the second ε to align j at the beginning. After that, process it using Eq. (2.21).

2.2.5 *Typical Applications of Index Operations*

As long as you master Eqs. (2.21), (2.13), and (2.19), you will be able to understand applications like the one below. The index operations introduced here represent a "standard routine" that frequently appears in subsequent discussions. Hence, it is intentionally included in the main text rather than relegated to a Chap. 9.

[19] Memorizing something meaningless, like a chant, is extremely difficult. However, once the meaning is understood, memorization becomes significantly easier.

2.2.5.1 Dot Product of Two Cross Products

The formula,

$$(\mathbf{A} \times \mathbf{B})\,(\mathbf{C} \times \mathbf{D}) = (\mathbf{A} \cdot \mathbf{C}) \cdot (\mathbf{B} \cdot \mathbf{D}) - (\mathbf{A} \cdot \mathbf{D}) \cdot (\mathbf{B} \cdot \mathbf{C}) \ , \tag{2.23}$$

known as Lagrange's identity, is often introduced in elementary derivations by writing out both sides explicitly in terms of components like (A_x, A_y, A_z) and verifying their equality. However, using what has been taught so far, the left-hand side can be computed directly to derive this result.

The left-hand side $(\mathbf{A} \times \mathbf{B})\,(\mathbf{C} \times \mathbf{D})$ is a "dot product of vectors". Using the notation convention for dot products in Eq. (2.9), we write,

$$(\mathbf{A} \times \mathbf{B})\,(\mathbf{C} \times \mathbf{D}) = (\mathbf{A} \times \mathbf{B})_l (\mathbf{C} \times \mathbf{D})_l \ .$$

Next, the l-component of each cross product can be expressed using Eq. (2.13) as

$$(\mathbf{A} \times \mathbf{B})_l = \varepsilon_{lst} A_s B_t \quad , \quad (\mathbf{C} \times \mathbf{D})_l = \varepsilon_{lmn} C_m D_n$$

using Eddington's symbols. When writing these out explicitly, it is crucial to assign indices carefully, keeping in mind that "l is fixed while (s, t) run over the summation index". If this is not done, errors may occur. As cautioned in the discussion of Eq. (2.10), assigning the same symbol to dummy variables under different summations can erroneously introduce connections between unrelated terms. It is essential to assign **distinct symbols to dummy variables in different summations**. Hence, if (s, t) are assigned to the summation over A and B, a different set (m, n) is assigned to the summation over C and D.

Thus, we have

$$(\mathbf{A} \times \mathbf{B})\,(\mathbf{C} \times \mathbf{D}) = (\mathbf{A} \times \mathbf{B})_l (\mathbf{C} \times \mathbf{D})_l = \varepsilon_{lst} A_s B_t \cdot \varepsilon_{lmn} C_m D_n \ .$$

Now, applying formula, Eq. (2.21), it leads to

$$\begin{aligned}
(\mathbf{A} \times \mathbf{B})\,(\mathbf{C} \times \mathbf{D}) &= \varepsilon_{lst} A_s B_t \cdot \varepsilon_{lmn} C_m D_n \\
&= \varepsilon_{lst} \varepsilon_{lmn} \cdot A_s B_t C_m D_n \\
&= (\delta_{sm}\delta_{tn} - \delta_{sn}\delta_{tm}) \cdot A_s B_t C_m D_n \\
&= \delta_{sm}\delta_{tn} \cdot A_s B_t C_m D_n - \delta_{sn}\delta_{tm} \cdot A_s B_t C_m D_n \ .
\end{aligned}$$

Although one could immediately arrive at Eq. (2.24) by applying formula, Eq. (2.21) directly, it is educational to explicitly follow the meaning of each term.[20] Writing out the summation for the first term explicitly, it leads to

$$\begin{aligned}\delta_{sm}\delta_{tn} \cdot A_s B_t C_m D_n &= \sum_s \sum_t \sum_m \sum_n \delta_{sm}\delta_{tn} \cdot A_s B_t C_m D_n \\ &= \sum_s \sum_t \left(A_s B_t \sum_m \sum_n \delta_{sm}\delta_{tn} \cdot C_m D_n \right) \\ &= \sum_s \sum_t A_s B_t C_s D_t \ .\end{aligned}$$

Here, the summation over (m, n) was performed first, leaving the summation over (s, t). The summation over (m, n) "contracts" the indices through the delta symbols, leaving $C_s D_t$. This process is often described as "contracting (m, n) through the deltas", and developing a sense for this phrasing is important.[21] The same process applies to the second term, resulting in

$$\begin{aligned}(\mathbf{A} \times \mathbf{B})(\mathbf{C} \times \mathbf{D}) &= \delta_{sm}\delta_{tn} \cdot A_s B_t C_m D_n - \delta_{sn}\delta_{tm} \cdot A_s B_t C_m D_n \\ &= A_s B_t C_s D_t - A_s B_t C_t D_s \ .\end{aligned} \tag{2.24}$$

Finally, substituting back the "aligned indices" under Einstein summation into the dot product notation, we obtain

$$\begin{aligned}(\mathbf{A} \times \mathbf{B})(\mathbf{C} \times \mathbf{D}) &= A_s B_t C_s D_t - A_s B_t C_t D_s \\ &= A_s C_s \cdot B_t D_t - A_s D_s \cdot B_t C_t \\ &= (\mathbf{A} \cdot \mathbf{C}) \cdot (\mathbf{B} \cdot \mathbf{D}) - (\mathbf{A} \cdot \mathbf{D}) \cdot (\mathbf{B} \cdot \mathbf{C}) \ ,\end{aligned} \tag{2.25}$$

arriving at Eq. (2.23).

2.2.5.2 Orthogonality in Cross Products

As learned in high school, the cross product $(\mathbf{A} \times \mathbf{B})$ is orthogonal to both $\mathbf{A}$ and $\mathbf{B}$. Thus,

$$\mathbf{A} \cdot (\mathbf{A} \times \mathbf{B}) = 0$$

[20] When you are newly accustomed to symbolic manipulation, you may sometimes lose sight of the original meaning. In such cases, it is essential to develop the habit of explicitly writing down the meanings and ensuring they align with the symbolic operations.

[21] In this case, we first "contracted" (m, n) to align with (s, t), but even if we had contracted (s, t) first, the result would have been the same. Therefore, it does not matter which is contracted first.

must be concluded. This can be demonstrated as follows. Using Eq. (2.9) for the dot product and Eq. (2.13) for the cross product, the left-hand side becomes

$$\mathbf{A} \cdot (\mathbf{A} \times \mathbf{B}) = A_l \cdot \varepsilon_{lst} A_s B_t = \varepsilon_{lst} A_l A_s \cdot B_t = \varepsilon_{tls} A_l A_s \cdot B_t \; .$$

However,

$$\varepsilon_{tls} A_l A_s = (\mathbf{A} \times \mathbf{A})_t = 0$$

and thus the statement is proven.

Here, we used the property that "the cross product of a vector with itself is zero". To delve deeper into why this holds, consider the relation:

$$(\mathbf{A} \times \mathbf{A})_t = \varepsilon_{tls} \cdot A_l A_s = 0 \; . \tag{2.26}$$

Recalling that the indices l and s are dummy index to be summed over, let us explicitly examine a fixed t and substitute specific indices a and b for l and s that yield non-zero terms:

$$\varepsilon_{tls} A_l A_s = \varepsilon_{tab} A_a A_b + \varepsilon_{tba} A_b A_a = (+1)\, A_a A_b + (-1)\, A_b A_a = 0 \; .$$

This result arises from the fact that $A_l A_s$ is symmetric (unchanged under exchange of indices), while ε_{tls} is antisymmetric (sign-reversing under exchange of indices). Therefore, the product vanishes.

In future encounters with structures like Eq. (2.26), strive to develop an operational understanding, recognizing that "since ε_{tls} is antisymmetric with respect to the exchange of (l, s), while $A_l A_s$ is symmetric, the result must be zero". This logic will reappear several times going forward.

2.3 Representation Theory of Linear Projections

Readers may recall their undergraduate experience where they were introduced to the concept of a "matrix", represented as

$$A = \begin{pmatrix} a_{11} & a_{12} & \cdots & a_{1N} \\ a_{21} & \ddots & & \vdots \\ \vdots & & & \\ a_{N1} & \cdots & & a_{NN} \end{pmatrix} ,$$

and taught operations such as matrix multiplication in a rote manner. This chapter aims to answer the question: what was the underlying purpose of that exercise?

Specifically, a matrix can be understood as a numerical representation of an abstract linear projection $\mathbf{y} = \hat{A} \cdot \mathbf{x}$ when expressed in terms of a given orthonormal basis $\{\mathbf{e}_j\}$.

The rules for matrix operations may be ingrained in many readers as part of their undergraduate exam preparation. Now, consider whether you could confidently answer the following questions if asked by a curious nephew or niece:

1. Why is matrix multiplication defined in such a peculiar way?
2. Why is the determinant defined in such a peculiar way?
3. Why can the inverse of a matrix be obtained through those specific operations?
4. Fundamentally, what necessitates the introduction of the determinant?

In this section, we aim to introduce concepts such as matrices, matrix multiplication, and diagonalisation in the context of "representing linear projections" as naturally as possible, while keeping these questions in mind.

2.3.1 *Bases and Projections*

A vector can be expressed in component form as:

$$\mathbf{x} = \begin{pmatrix} x_1 \\ x_2 \\ \vdots \\ x_N \end{pmatrix} = \sum_l x_l \cdot \mathbf{e}_l \; . \tag{2.27}$$

By taking the dot product of both sides with $\mathbf{x}$ and $\mathbf{e}_j$, and using the orthogonality of the basis (2.12), we have

$$\left(\mathbf{x} \cdot \mathbf{e}_j\right) = \sum_l x_l \cdot \left(\mathbf{e}_l \cdot \mathbf{e}_j\right) = \sum_l x_l \cdot \delta_{lj} = x_j \; . \tag{2.28}$$

Thus, we obtain:

$$x_j = \left(\mathbf{x}, \mathbf{e}_j\right) \tag{2.29}$$

This result, obtained operationally here, is something taught in high school: each component of a vector is the **projection** of the vector onto each basis vector. The operation of "taking the dot product with $\mathbf{e}_j$" geometrically extracts the length of the shadow along the $\mathbf{e}_j$ axis. Reflect on this meaning and practise deriving Eq. (2.29) directly.

Substituting Eq. (2.29) into Eq. (2.27), we have

$$\mathbf{x} = \sum_l (\mathbf{x}, \mathbf{e}_l) \cdot \mathbf{e}_l \ . \tag{2.30}$$

In component form, this becomes:

$$x_j = (\mathbf{x} \cdot \mathbf{e}_l) \cdot [\mathbf{e}_l]_j = (x_m \cdot [\mathbf{e}_l]_m) \cdot [\mathbf{e}_l]_j = x_m \cdot [\mathbf{e}_l]_m \cdot [\mathbf{e}_l]_j \tag{2.31}$$

Here, $[\mathbf{e}_k]_j$ denotes the j-th component of the k-th basis vector $\mathbf{e}_k$. When deriving the first equality, note that $(\mathbf{x} \cdot \mathbf{e}_l)$ is a scalar, while $\mathbf{e}_l$ is a vector. The left-hand side of Eq. (2.30) is a vector $\mathbf{x}$ because the vector nature is carried by $\mathbf{e}_l$ on the right-hand side. Therefore, "taking the j-th component" involves taking x_j on the left-hand side and $[\mathbf{e}_l]_j$ on the right-hand side.

Considering the summation over m in the rightmost term of Eq. (2.31), for the equation to hold, the following condition must be satisfied:

$$[\mathbf{e}_k]_m [\mathbf{e}_k]_j = \delta_{mj}. \ .$$

This ensures that only x_j from the summation over x_m in the rightmost term is picked out, making both sides equal. This requirement is called the **completeness of the basis**.

It is worth emphasizing again that some students often confuse basis completeness with basis orthogonality, so take special care to distinguish between the two. Basis orthogonality is expressed as:

$$\mathbf{e}_m \cdot \mathbf{e}_j = \delta_{mj}, \ ,$$

or equivalently:

$$[\mathbf{e}_m]_k \left[\mathbf{e}_j\right]_k = \delta_{mj}. \ .$$

To restate and juxtapose,[22]

$$\begin{aligned} [\mathbf{e}_m]_k \left[\mathbf{e}_j\right]_k &= \delta_{mj} \quad \cdots \text{(orthogonality)} \ , \\ [\mathbf{e}_k]_m [\mathbf{e}_k]_j &= \delta_{mj} \quad \cdots \text{(completeness)} \ . \end{aligned} \tag{2.32}$$

You may find the term "completeness" perplexing. While "orthogonality" is intuitive, why is

$$[\mathbf{e}_k]_m [\mathbf{e}_k]_j = \delta_{mj}$$

[22] Avoid skimming over this part as "confusing and difficult". Write it out yourself a few times to fully grasp it. As with anything, the beginning is crucial; take your time to carefully work through the initial concepts. This book is structured to provide thorough explanations early on rather than relegating them to exercises or Chap. 9.

called completeness (meaning "fully equipped")? The key lies in recognizing that the completeness condition arises from the requirement that Eq. (2.30) holds. This may seem obvious, but Eq. (2.30) asserts that "after projecting $\mathbf{x}$ into its components and reconstructing the vector from those components, the original $\mathbf{x}$ is fully recovered without any loss of information". This is not necessarily guaranteed in general contexts. A common analogy is that "schools profile students by their grades in different subjects, but those grades alone cannot fully reconstruct the original individual".

Thus, Eq. (2.30) assumes that the basis set $\{\mathbf{e}_j\}$ fully represents the space without omissions and is therefore **complete**. The resulting Eq. (2.32) embodies what we call "basis completeness".

Classification of Spaces

For elementary vector spaces, operations like "taking projections" and "reconstructing from projections to recover the original" are often self-evident or taken for granted. However, when dealing with more general spaces, such as "data spaces", it is crucial to explicitly consider whether these operations are feasible or guaranteed. A space where component decomposition via projection is possible within certain limits is called a **Banach space**, while a space where complete reconstruction from the projected components is guaranteed is called a **Hilbert space**. In quantum mechanics, functions are treated as "infinite-dimensional vectors", and the logic is developed with the assumption that the original function can be expressed as a sum of "components multiplied by basis functions", as shown in Eq. (2.27). The foundation of this reasoning lies in the premise that complete reconstruction from the projected components is ensured. When studying quantum mechanics, students often encounter the term "Hilbert space" introduced suddenly by instructors without much explanation, leading to confusion. This arises from the considerations outlined above.

2.3.2 *Representation of Linear Projections*

Let us consider a vector $\mathbf{x}$, which undergoes a projection $\hat{A}$ to produce another vector $\mathbf{y}$, interpreted as being "mapped" (projected) to a new position. We express this mapping as $\mathbf{y} = \hat{A} \cdot \mathbf{x}$. Note that at this stage, nothing has been said about $\hat{A}$ being a matrix or square array. Here, $\hat{A}$ is assumed to be a linear projection, satisfying,

$$\hat{A} \cdot (\lambda \cdot \mathbf{x} + \mu \cdot \mathbf{y}) = \lambda \cdot \left(\hat{A} \cdot \mathbf{x}\right) + \mu \cdot \left(\hat{A} \cdot \mathbf{y}\right) .. \tag{2.33}$$

At this point, we introduce the idea that the mapped vector $\mathbf{y} = \hat{A}\mathbf{x}$ "remains in the same **village** as **x**."[23] This requirement is mathematically represented by the notion that "$\hat{A}$ is represented by a numerical square array."[24] For beginners who might find this paragraph unclear, reaching a point where this makes sense marks the goal of understanding. After completing this section, it is recommended to return and re-read this part to check your comprehension.

The next step is to mathematically represent the concept that "the mapped vector **y** remains in the same village as the original vector **x**". This is expressed as "**y** is represented in terms of the same basis set $\{\mathbf{e}_j\}$ that constitutes **x**". Substituting,

$$\mathbf{x} = \sum_l x_l \cdot \mathbf{e}_l \ ,$$

into $\mathbf{y} = \hat{A} \cdot \mathbf{x}$, and using the linearity in Eq. (2.33), we obtain,

$$\mathbf{y} = \sum_l x_l \cdot \left(\hat{A} \cdot \mathbf{e}_l\right). \tag{2.34}$$

Thus, **y** is expressed as a superposition of the projections $\hat{A} \cdot \mathbf{e}_l$ of each basis vector. Consequently, the requirement that "**y** remains in the same village" can be rephrased as the requirement that each projection $\hat{A} \cdot \mathbf{e}_l$ can be expanded as

$$\hat{A} \cdot \mathbf{e}_j = \sum_m c_m \cdot \mathbf{e}_m, \tag{2.35}$$

using the same basis set $\{\mathbf{e}_j\}$. As Eq. (2.34) suggests, the projection rule for any $\mathbf{x} \to \mathbf{y}$ is governed by the projection rule for $\hat{A} \cdot \mathbf{e}_l$. This leads to the understanding that $\hat{A}$'s nature can be captured by the projection rule for the basis vectors, Eq. (2.35).

Equation (2.35) is an incomplete representation, as the expansion coefficients differ for each basis indexed by j. To reflect the index j of the transformed basis in the coefficients, we adopt the notation:

$$\hat{A} \cdot \mathbf{e}_j = \sum_m a_m^{(j)} \cdot \mathbf{e}_m = \sum_k a_{mj} \cdot \mathbf{e}_m. \tag{2.36}$$

[23] For example, in a mapping where rational numbers are transformed into irrational numbers, the image would "moving out to another village". This aspect reflects the underlying need for a precise mathematical framework in real analysis, beyond mere intuition.

[24] This is often referred to as the "representation matrix of the operator $\hat{A}$". The fact that group theory often culminates in "matrix operations" is fundamentally tied to this idea of "the projection remaining in the same village".

Here, $a_m^{(j)} = a_{mj}$, and the convention of placing j as the latter subscript in a_{mj} aligns with standard notations in other references. This perspective that "$\hat{A}$'s nature is captured by the projection rule for the basis vectors Eq. (2.36)" essentially equates $\hat{A}$ with:

$$\hat{A} \quad \longleftrightarrow \quad \{a_{mj}\}.$$

The set $\{a_{mj}\}$ is the matrix:

$$\hat{A} \quad \longleftrightarrow \quad A = \begin{pmatrix} a_{11} & a_{12} & \cdots & a_{1N} \\ a_{21} & & \ddots & \vdots \\ \vdots & & & \\ a_{N1} & \cdots & & a_{NN} \end{pmatrix},$$

which corresponds to the numerical square array we are familiar with.

Following the approach in Eq. (2.28), taking the dot product of both sides of Eq. (2.36) with $\mathbf{e}_k$, we find:

$$\left(\mathbf{e}_k \cdot \left(\hat{A} \cdot \mathbf{e}_j\right)\right) = \left(\mathbf{e}_k \cdot \left(\sum_m a_{mj} \cdot \mathbf{e}_m\right)\right) = \sum_m a_{mj} \cdot (\mathbf{e}_k \cdot \mathbf{e}_m)$$
$$= \sum_m a_{mj} \cdot \delta_{km} = a_{kj}.$$

Thus, we obtain:

$$a_{kj} = \left(\mathbf{e}_k, \hat{A} \cdot \mathbf{e}_j\right). \tag{2.37}$$

In summary, the nature of an abstract linear projection $\hat{A}$ can be captured by:

$$\hat{A} \cdot \mathbf{e}_j = \sum_m a_{mj} \cdot \mathbf{e}_m, \tag{2.38}$$

which allows $\hat{A}$ to be identified with the matrix:

$$\hat{A} \quad \longleftrightarrow \quad A = \begin{pmatrix} a_{11} & a_{12} & \cdots & a_{1N} \\ a_{21} & & \ddots & \vdots \\ \vdots & & & \\ a_{N1} & \cdots & & a_{NN} \end{pmatrix}.$$

The elements of this matrix can be determined by calculating:

$$a_{kj} = \left(\mathbf{e}_k \cdot \hat{A} \cdot \mathbf{e}_j\right),$$

a process referred to as the **extraction of matrix elements** .

2.3.3 Operation Rules for Representations of Linear Projections

Equation (2.38) forms the fundamental connection between a linear projection $\hat{A}$ and its representation matrix:

$$a_{kj} = \left(\mathbf{e}_k \cdot \hat{A} \cdot \mathbf{e}_j\right).$$

Using this, we can derive the following rules:

- How are y_j and x_j related in the mapping $\mathbf{y} = \hat{A} \cdot \mathbf{x}$, expressed using $\{a_{kj}\}$?
- How are the representation matrices c_{ij}, b_{ij}, and a_{ij} related in a composite projection $\hat{C} = \hat{B} \cdot \hat{A}$?

These rules correspond to the operations for "vector-matrix multiplication" and "matrix-matrix multiplication" introduced in undergraduate studies. Let us explore them in detail.

To establish the relationship between components y_j and x_j in $\mathbf{y} = \hat{A}\cdot\mathbf{x}$, consider the following. Substituting Eq. (2.38) for $\hat{A} \cdot \mathbf{x}$:

$$\hat{A} \cdot \mathbf{x} = \sum_l x_l \cdot \left(\hat{A} \cdot \mathbf{e}_l\right) = \sum_l x_l \cdot \left(\sum_m a_{ml} \cdot \mathbf{e}_m\right) = \sum_{m,l} a_{ml} x_l \cdot \mathbf{e}_m. \quad (2.39)$$

Equating this to:

$$\mathbf{y} = \sum_m y_m \cdot \mathbf{e}_m,$$

and comparing the coefficients of $\mathbf{e}_m$, we obtain:

$$y_m = \sum_l a_{m\underline{l}} \cdot x_{\underline{l}}. \quad (2.40)$$

Writing this explicitly for each component:

$$
\begin{aligned}
y_1 &= a_{1\underline{1}}x_{\underline{1}} + a_{1\underline{2}}x_{\underline{2}} + a_{1\underline{3}}x_{\underline{3}} + \cdots \\
y_2 &= a_{2\underline{1}}x_{\underline{1}} + a_{2\underline{2}}x_{\underline{2}} + a_{2\underline{2}}x_{\underline{3}} + \cdots \\
&\quad \ldots
\end{aligned}
$$

This corresponds to the familiar **vector-matrix product rule** learned in undergraduate studies:

$$
\begin{pmatrix} y_1 \\ y_2 \\ \vdots \\ y_N \end{pmatrix} = \begin{pmatrix} a_{11} & a_{12} & \cdots & a_{1N} \\ a_{21} & \ddots & & \vdots \\ \vdots & & & \\ a_{N1} & \cdots & & a_{NN} \end{pmatrix} \begin{pmatrix} x_1 \\ x_2 \\ \vdots \\ x_N \end{pmatrix}. \tag{2.41}
$$

It is crucial to note that this operation rule stems from the projection rule for each basis vector given in Eq. (2.38).

In summary, the following applies: When performing or calculating the representation of an abstract linear operation $\mathbf{y} = \hat{A} \cdot \mathbf{x}$ in terms of specific components on an orthonormal basis $\{\mathbf{e}_j\}$, the first step is to represent the vector $\mathbf{x} = \sum_k x_k \cdot \mathbf{e}_k$ as:

$$
\begin{pmatrix} x_1 \\ x_2 \\ \vdots \\ x_N \end{pmatrix} \leftrightarrow \mathbf{x} \quad , \quad \begin{pmatrix} y_1 \\ y_2 \\ \vdots \\ y_N \end{pmatrix} \leftrightarrow \mathbf{y}.
$$

Then, identifying the linear operator $\hat{A}$ with the square array constructed using the values:

$$
a_{kj} = \left(\mathbf{e}_k \cdot \hat{A} \cdot \mathbf{e}_j \right),
$$

we write:

$$
A = \{a_{kj}\} = \begin{pmatrix} a_{11} & a_{12} & \cdots & a_{1N} \\ a_{21} & \ddots & & \vdots \\ \vdots & & & \\ a_{N1} & \cdots & & a_{NN} \end{pmatrix} \leftrightarrow \hat{A}.
$$

The components of **y** can then be obtained from the components of **x** as:

$$y_j = a_{j\underline{l}} \cdot x_{\underline{l}}, \tag{2.42}$$

where we have removed the summation symbol following Einstein's convention and replaced the index m with j from Eq. (2.40). The matrix-vector calculation rules introduced in high school, such as Eq. (2.41), can now be compactly remembered in the form of Eq. (2.42). Take note of the contraction of the index l. The index l in a_{jl} contracts with x_l, leaving a single surviving index j that represents the resulting vector. Here, l serves as a dummy index, while j is the "surviving index" that forms the vector.

While it is possible to "skim through" the above, it is highly recommended to rewrite the index manipulations yourself and try explaining them to others. Doing so will help you truly master the handling of indices. For instance, by writing out the details yourself, you may notice the difference in the role of the dummy index m in the following equations:

$$\hat{A} \cdot \mathbf{e}_j = \sum_m a_{mj} \cdot \mathbf{e}_m \quad \text{(Projection rule for the basis)}, \tag{2.43}$$

and:

$$y_j = \sum_m a_{jm} \cdot x_m \quad \text{(Projection rule for the components)}. \tag{2.44}$$

This distinction corresponds to the difference between contravariant and covariant quantities [4]. It is recommended to aim for a level of familiarity where you can recite such index manipulations. The reason is that the "framework of vector spaces" discussed here will reappear in various forms across different contexts in the future. For instance, when delving into applications related to the representation theory of groups, you will encounter structures almost identical to matrix diagonalisation (or simplification) in the form of reduction operations [1].[25] This is because the foundational idea in group representation theory, "remaining in the same village", is analogous to the concept of vector space representations.

Seemingly Tedious Foundations Are Actually Essential
When studying "mathematics" in undergraduate courses, you may encounter topics like the definition of vector spaces, injective and surjective mappings, isomorphisms, and set or topology concepts. These subjects, which often lack

(continued)

[25] Refer to the Column in Sect. 2.3.9 for further details.

concrete imagery, appear early in the curriculum and may seem confusing or difficult to appreciate. You might wonder why mathematicians place so much emphasis on these "tedious topics". In reality, various mathematical methodologies that emerged from different specific contexts and historical developments have been reinterpreted and unified under common concepts like sets, topology, and vector spaces. It has become evident that a shared logical structure underpins these diverse methods. The seemingly dry subjects introduced in the first year serve as a foundational "common language" for appreciating this intellectual framework. For this reason, they are given significant emphasis, despite appearing uninteresting at first glance.

2.3.4 *Operational Rules for Representations of Composite Mappings*

Consider the superposition of mappings shown in Fig. 2.2:

$$\mathbf{z} = \hat{B} \cdot \mathbf{y} = \hat{B} \cdot \left(\hat{A} \cdot \mathbf{x}\right) =: \hat{C} \cdot \mathbf{x}.$$

Recalling Eq. (2.42), this can be expressed in terms of components as:

$$z_j = b_{jk} y_k = b_{jk} \left(a_{kl} x_l\right), \quad \text{and} \quad z_j = c_{jl} x_l.$$

Comparing the two expressions above yields:

$$c_{jl} = b_{jk} a_{kl}, \tag{2.45}$$

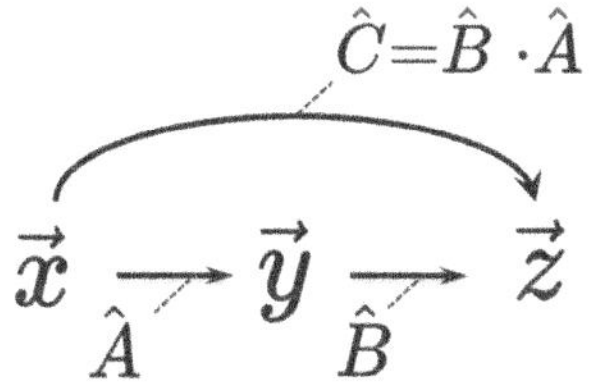

Fig. 2.2 Diagram illustrating the composite mapping $\hat{B} \cdot \hat{A}$. When $\mathbf{x}$ is mapped to $\mathbf{y}$ by $\hat{A}$, and subsequently $\mathbf{y}$ is mapped to $\mathbf{z}$ by $\hat{C}$, we can write: $\mathbf{z} = \left(\hat{B} \cdot \hat{A}\right) \cdot \mathbf{x}$. If we express this as $\mathbf{z} = \hat{C} \cdot \mathbf{x}$, we consider how the representation matrix of $\hat{C}$ is described in terms of the representation matrices of $\hat{B}$ and $\hat{A}$

which defines the **projection rule for the matrix components** corresponding to the composition of mappings $\hat{C} = \hat{B} \cdot \hat{A}$. Here, it is noted to develop an intuitive understanding that the index k is a dummy variable summing over pairs and "contracted", while the surviving indices j and l are reflected in the left-hand side.

Expanding Eq. (2.45) explicitly, we obtain:[26]

$$\begin{aligned} c_{11} &= b_{1\underline{1}}a_{\underline{1}1} + b_{1\underline{2}}a_{\underline{2}1} + \cdots \\ c_{12} &= b_{1\underline{1}}a_{\underline{1}2} + b_{1\underline{2}}a_{\underline{2}2} + \cdots \\ &\cdots \\ c_{21} &= b_{2\underline{1}}a_{\underline{1}1} + b_{2\underline{2}}a_{\underline{2}1} + \cdots . \\ &\cdots \end{aligned}$$

This corresponds to the familiar **matrix multiplication rule** learned in undergraduate studies:

$$\begin{pmatrix} c_{11} & c_{12} & \cdots & c_{1N} \\ c_{21} & \ddots & & \\ \vdots & & & \\ c_{N1} & \cdots & & c_{NN} \end{pmatrix} = \begin{pmatrix} b_{11} & b_{12} & \cdots & b_{1N} \\ b_{21} & \ddots & & \\ \vdots & & & \\ b_{N1} & \cdots & & b_{NN} \end{pmatrix} \begin{pmatrix} a_{11} & a_{12} & \cdots & a_{1N} \\ a_{21} & \ddots & & \\ \vdots & & & \\ a_{N1} & \cdots & & a_{NN} \end{pmatrix} .$$

Summarizing the discussion thus far, we can reiterate the following: For vector-matrix and matrix-matrix products, the component-wise representations are, respectively:

$$y_j = a_{j\underline{m}}x_{\underline{m}} \quad , \quad c_{ij} = b_{i\underline{m}}a_{\underline{m}j} \; . \tag{2.46}$$

Matrix components themselves are defined from the linear mapping $\hat{A}$ as:

$$a_{kj} = \left(\mathbf{e}_k, \hat{A} \cdot \mathbf{e}_j\right) \; , \tag{2.47}$$

where the projection of the basis system is expressed as:

$$\hat{A} \cdot \mathbf{e}_j = a_{\underline{m}j} \cdot \mathbf{e}_{\underline{m}} . \tag{2.48}$$

[26] It is important to repeatedly practice such explicit expansions to fully grasp the meaning of indices.

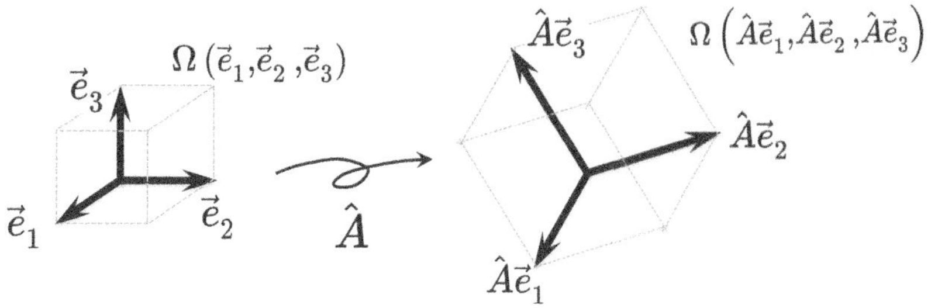

Fig. 2.3 The volume of the parallelepiped formed by the basis vectors of the coordinate system, $\{\mathbf{e}_j\}_{j=1}^N$, is denoted as $\Omega^{(N)}(\mathbf{e}_1, \cdots, \mathbf{e}_N)$. When this parallelepiped is transformed by $\hat{A}$, the resulting "new frame" forms a parallelepiped whose volume is expressed as $\Omega^{(N)}(\mathbf{e}'_1, \cdots, \mathbf{e}'_N)$. The scaling factor of this volume is defined as the determinant $|A|$

2.3.5 Introduction to Determinants

Since a linear map can be represented by a square matrix $\left\{a_{ij} = \left(\mathbf{e}_i, \hat{A}\mathbf{e}_j\right)\right\}$, it is natural to consider a metric for its absolute magnitude. The "magnitude" $|A|$ of the linear map $\hat{A}$ can naturally be understood as the "scaling factor of projection". Specifically, the metric $|A|$ is defined as the factor by which the "unit volume" in N-dimensional space expands when projected by $\hat{A}$, as shown in Fig. 2.3. This metric, $|A| = \det A$, is referred to as the **determinant** of the matrix A.

The "unit volume in N-dimensional space" is defined as the volume of the parallelepiped spanned by the basis vectors $\{\mathbf{e}_j\}_{j=1}^N$, denoted as $\Omega^{(N)}(\mathbf{e}_1, \cdots, \mathbf{e}_N)$. When this parallelepiped is projected by $\hat{A}$, the resulting structure is spanned by $\left\{\mathbf{e}'_j = \hat{A}\mathbf{e}_j\right\}$, and its volume is given by $\Omega^{(N)}\left(\mathbf{e}'_1, \cdots, \mathbf{e}'_N\right)$. Thus, $|A|$ can be expressed as:

$$\Omega^{(N)}\left(\mathbf{e}'_1, \cdots, \mathbf{e}'_N\right) = |A| \cdot \Omega^{(N)}(\mathbf{e}_1, \cdots, \mathbf{e}_N). \tag{2.49}$$

By defining the determinant as the "scaling factor associated with the mapping", it follows naturally that for composite mappings:

$$\det[AB] = \det[A] \cdot \det[B]. \tag{2.50}$$

To formalize this idea further, let us consider how the function $\Omega^{(N)}(\mathbf{v}_1, \cdots, \mathbf{v}_N)$, which gives the volume of the parallelepiped spanned by N vectors $\{\mathbf{v}_j\}_{j=1}^N$ in N-dimensional space, behaves. For simplicity, consider the two-dimensional area $\Omega^{(2)}(\mathbf{v}_1, \mathbf{v}_2)$ as an example. For this, we can confirm the bilinearity,

$$\Omega^{(2)}\left(\mathbf{v}_1 + \mathbf{v}'_1, \mathbf{v}_2\right) = \Omega^{(2)}(\mathbf{v}_1, \mathbf{v}_2) + \Omega^{(2)}\left(\mathbf{v}'_1, \mathbf{v}_2\right)$$
$$\Omega^{(2)}\left(\mathbf{v}_1, \mathbf{v}_2 + \mathbf{v}'_2\right) = \Omega^{(2)}(\mathbf{v}_1, \mathbf{v}_2) + \Omega^{(2)}\left(\mathbf{v}_1, \mathbf{v}'_2\right)$$

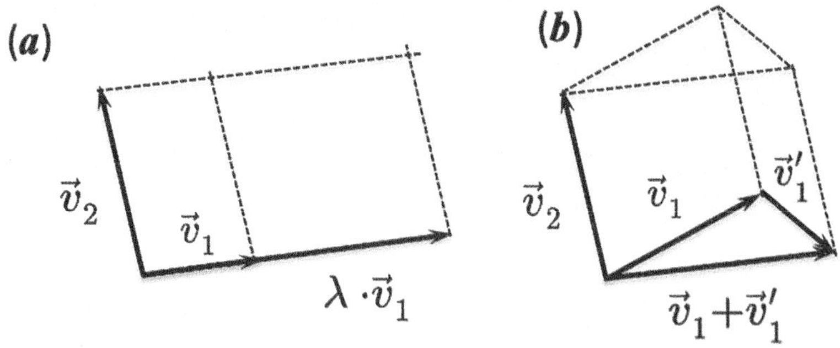

Fig. 2.4 (**a**) Compared to the area formed by v_2 and v_1, the area formed by v_2 and $\lambda \cdot v_1$ becomes λ times the original area. (**b**) Also, by considering which parts of the figure correspond to the area formed by the vector sum $\mathbf{v}_1 + \mathbf{v}'_1$ together with $\mathbf{v}_2$, and the areas formed by $\mathbf{v}_1$ and $\mathbf{v}'_1$ each paired with $\mathbf{v}_2$, we can verify that the bilinearity described in Eq. (2.51) indeed holds

$$\Omega^{(2)}(\mathbf{v}_1, \lambda\mathbf{v}_2) = \lambda \cdot \Omega^{(2)}(\mathbf{v}_1, \mathbf{v}_2)$$
$$\cdots, \tag{2.51}$$

as illustrated in Fig. 2.4.

Furthermore, an important property is that:

$$\Omega^{(2)}(\mathbf{v}, \mathbf{v}) = 0\,, \tag{2.52}$$

since the area spanned by the same vector is zero. From this property, we derive the antisymmetry as

$$0 = \Omega^{(2)}(\mathbf{v} + \mathbf{w}, \mathbf{v} + \mathbf{w}) = \Omega^{(2)}(\mathbf{v}, \mathbf{w}) + \Omega^{(2)}(\mathbf{w}, \mathbf{v})$$
$$\therefore \quad \Omega^{(2)}(\mathbf{v}, \mathbf{w}) = -\Omega^{(2)}(\mathbf{w}, \mathbf{v})\,. \tag{2.53}$$

In three dimensions, the volume $\Omega^{(3)}(\mathbf{v}_1, \mathbf{v}_2, \mathbf{v}_3)$ satisfies the same property: if any two vectors are identical, the parallelepiped collapses, and the volume becomes zero. Using the same derivation as Eq. (2.53), we obtain the **antisymmetry with respect to permutations**:

$$\Omega^{(3)}\left(\underline{\mathbf{v}_1}, \underline{\mathbf{v}_2}, \mathbf{v}_3\right) = -\Omega^{(3)}\left(\underline{\mathbf{v}_2}, \underline{\mathbf{v}_1}, \mathbf{v}_3\right),$$
$$\Omega^{(3)}\left(\mathbf{v}_2, \underline{\mathbf{v}_1}, \underline{\mathbf{v}_3}\right) = -\Omega^{(3)}\left(\mathbf{v}_2, \underline{\mathbf{v}_3}, \underline{\mathbf{v}_1}\right),$$
$$\cdots$$

where the underlined vectors have been swapped. Continuing these permutations, it leads to

$$\begin{aligned}\Omega^{(3)}(\mathbf{v}_1,\mathbf{v}_2,\mathbf{v}_3) &= (-)\cdot\Omega^{(3)}(\mathbf{v}_2,\mathbf{v}_1,\mathbf{v}_3)\\ &= (-)^2\cdot\Omega^{(3)}(\mathbf{v}_2,\mathbf{v}_3,\mathbf{v}_1)\\ &= (-)^3\cdot\Omega^{(3)}(\mathbf{v}_3,\mathbf{v}_2,\mathbf{v}_1)\\ &= (-)^4\cdot\Omega^{(3)}(\mathbf{v}_3,\mathbf{v}_1,\mathbf{v}_2)\\ &= \cdots\,. \end{aligned} \tag{2.54}$$

2.3.6 Permutation Symbols and Their Signs

Equation (2.54) can be compactly expressed as:

$$\Omega^{(3)}\left(\mathbf{v}_{P_1},\mathbf{v}_{P_2},\mathbf{v}_{P_3}\right) = (-)^P\cdot\Omega^{(3)}(\mathbf{v}_1,\mathbf{v}_2,\mathbf{v}_3)\,, \tag{2.55}$$

but many students struggle with this notation, so we will explain it in detail.

The arguments on the right-hand side of Eq. (2.54) take the general form $\Omega^{(3)}\left(\mathbf{v}_i,\mathbf{v}_j,\mathbf{v}_k\right)$. All the indices (i,j,k) represent permutations of $(1,2,3)$, as such these are written as P_1, P_2, P_3 (P stands for "permutation"), and the general arguments are now expressed as $\Omega^{(3)}\left(\mathbf{v}_{P_1},\mathbf{v}_{P_2},\mathbf{v}_{P_3}\right)$. For example, a normal summation would be written as:

$$\begin{aligned} I &= \sum_{i,j,k} F\left(x_i,x_j,x_k\right)\\ &= F(x_1,x_1,x_1) + F(x_1,x_1,x_2) + \cdots\,, \end{aligned}$$

where all combinations of $(i,j,k) = (1,1,1), (1,1,2), \cdots$ are included, resulting in $3^3 = 27$ terms. However, for the case of permutations:

$$\begin{aligned} I &= \sum_{P_1,P_2,P_3} F\left(x_{P_1},x_{P_2},x_{P_3}\right) = \sum_P F\left(x_{P_1},x_{P_2},x_{P_3}\right)\\ &= F(x_1,x_2,x_3) + F(x_1,x_3,x_2) + \cdots\,, \end{aligned} \tag{2.56}$$

the indices in the arguments are limited to permutations of $(1,2,3)$, resulting in only $3! = 6$ terms [as shown in the left diagram of Fig. 2.5]. As indicated in the equation, the summation over P_1, P_2, P_3 is often simply written as a summation over P.

Next, we introduce the notation for the "sign of a permutation", $(-)^P$. This symbol represents the sign based on the rule of "applying a negative sign for each swap" as shown in Fig. 2.5 ($(-)^3$, etc.). The sign appearing in Eq. (2.54)

```
1   2   3                      1   2   3   4
P1  P2  P3                     P1  P2  P3  P4
----------------               ---------------
1   2   3   (+)                1   2   3   4   (+)
1   3   2   (-)                1   2   4   3   (-)
3   1   2   (+)=(-)^2          1   4   2   3   (+)=(-)^2
3   2   1   (-)=(-)^3          4   1   2   3   (-)=(-)^3
2   3   1   (+)=(-)^4          4   1   3   2   (+)=(-)^4
2   1   3   (-)=(-)^5          4   3   1   2   (-)=(-)^5
                               3   4   1   2   (+)=(-)^6
                               3   4   2   1   (-)=(-)^7
                               ...
```

Fig. 2.5 Permutations of three elements (left) and four elements (right). In each case, a new permutation is created by swapping the two underlined elements and moving to the next row

corresponds to $(-)^P$, which depends on how many swaps are required to transform $(1, 2, 3)$ into the given permutation. This allows Eq. (2.55) to be written as shown.

For two-dimensional areas, the bilinearity was verified in Fig. 2.4. Similarly, for higher-dimensional volume functions $\Omega^{(N)}(\mathbf{v}_1, \mathbf{v}_2, \cdots, \mathbf{v}_N)$, bilinearity holds as well. Furthermore, if any two vectors in the arguments are identical, the frame collapses, and the volume becomes zero. Using the same derivation as Eq. (2.53), we obtain the antisymmetry under swaps $\mathbf{v}_i \leftrightarrow \mathbf{v}_j$:

$$\Omega^{(N)}\left(\cdots, \mathbf{v}_i, \cdots, \mathbf{v}_j, \cdots\right) = -\Omega^{(N)}\left(\cdots, \mathbf{v}_j, \cdots, \mathbf{v}_i, \cdots\right).$$

Considering that each swap introduces a negative sign, as in Eq. (2.54), the permutation $\left(\mathbf{v}_{P_1}, \mathbf{v}_{P_2}, \cdots, \mathbf{v}_{P_N}\right)$ with respect to the standard order $(\mathbf{v}_1, \mathbf{v}_2, \cdots, \mathbf{v}_N)$ can be related via the permutation sign $(-)^P$:

$$\Omega^{(N)}\left(\mathbf{v}_{P_1}, \mathbf{v}_{P_2}, \cdots, \mathbf{v}_{P_N}\right) = (-)^P \cdot \Omega^{(N)}(\mathbf{v}_1, \mathbf{v}_2, \cdots, \mathbf{v}_N). \tag{2.57}$$

2.3.7 *Expression of the Determinant*

Building upon the above preparation, we revisit Eq. (2.49) to derive the expression for the determinant as follows: Given that the basis vectors $\left\{\mathbf{e}_j\right\}_{j=1}^N$ in an N-dimensional space are projected under a linear projection $\hat{A}$ into $\left\{\mathbf{e}'_j\right\}_{j=1}^N$,

$$\left\{\mathbf{e}'_j = A \cdot \mathbf{e}_j\right\}_{j=1}^N, \tag{2.58}$$

the ratio of the volume elements is defined as follows, representing the determinant of the linear projection $\hat{A}$:

$$\Omega^{(N)}\left(\mathbf{e}_1',\cdots,\mathbf{e}_N'\right)=(\det A)\cdot\Omega^{(N)}\left(\mathbf{e}_1,\cdots,\mathbf{e}_N\right).$$

From the projection rule, Eq. (2.48),

$$\mathbf{e}_j{}'=\hat{A}\cdot\mathbf{e}_j=a_{mj}\cdot\mathbf{e}_m. \tag{2.59}$$

Substituting this leads to

$$\Omega^{(N)}\left(\mathbf{e}_1',\mathbf{e}_2'\cdots,\mathbf{e}_N'\right)=\Omega^{(N)}\left(a_{m_1,1}\cdot\mathbf{e}_{m_1};\ a_{m_2,2}\cdot\mathbf{e}_{m_2};\ \cdots;\ a_{m_N,N}\cdot\mathbf{e}_{m_N}\right),$$

where we used semicolons instead of commas just for better readability. The dummy index m in Eq. (2.59) appears for each $\mathbf{e}_j$, but since the dummy index for different j are unrelated, they must not overlap. To emphasise this distinction, we write them as m_j.[27]

Due to bilinearity, each coefficient $a_{m_j,j}$ can be factored out of $\Omega^{(N)}$, and explicitly writing the summation over all m_j, we obtain:

$$\Omega^{(N)}\left(\mathbf{e}_1',\mathbf{e}_2'\cdots,\mathbf{e}_N'\right)=\sum_{m_1,m_2,\cdots,m_N}a_{m_1,1}\cdot a_{m_2,2}\cdots a_{m_N,N}\cdot\Omega^{(N)}\left(\mathbf{e}_{m_1},\mathbf{e}_{m_2},\cdots,\mathbf{e}_{m_N}\right).$$

Since the volume function vanishes if any two vectors in its argument $\left(\mathbf{e}_{m_1},\mathbf{e}_{m_2},\cdots,\mathbf{e}_{m_N}\right)$ are identical, only the terms where the integers assigned to $\{m_1,m_2,\cdots,m_N\}$ are distinct (i.e., permutations) contribute. Rewriting $\{m_1,m_2,\cdots,m_N\}$ as the permutation $\{p_1,p_2,\cdots,p_N\}$, we have:

$$\Omega^{(N)}\left(\mathbf{e}_1',\mathbf{e}_2'\cdots,\mathbf{e}_N'\right)=\sum_{p_1,p_2,\cdots,p_N}a_{p_1,1}\cdot a_{p_2,2}\cdots a_{p_N,N}\cdot\Omega^{(N)}\left(\mathbf{e}_{p_1},\mathbf{e}_{p_2},\cdots,\mathbf{e}_{p_N}\right).$$

Substituting Eq. (2.57):

$$\Omega^{(N)}\left(\mathbf{e}_{p_1},\mathbf{e}_{p_2},\cdots,\mathbf{e}_{p_N}\right)=(-)^P\cdot\Omega^{(N)}\left(\mathbf{e}_1,\mathbf{e}_2,\cdots,\mathbf{e}_N\right),$$

we obtain:

$$\Omega^{(N)}\left(\mathbf{e}_1',\mathbf{e}_2'\cdots,\mathbf{e}_N'\right)=\sum_{p_1,p_2,\cdots,p_N}(-)^P a_{p_1,1}\cdot a_{p_2,2}\cdots a_{p_N,N}\cdot\Omega^{(N)}\times\left(\mathbf{e}_1,\mathbf{e}_2,\cdots,\mathbf{e}_N\right).$$

[27] Recall the discussion in Eq. (2.10).

Thus, the proportionality factor is given as:

$$\det A = \sum_P (-)^P a_{P_1,1} \cdot a_{P_2,2} \cdots a_{P_N,N}. \tag{2.60}$$

This expression provides an explicit method for computing the determinant from the matrix representation of the linear projection $\hat{A}$ [Eq. (2.47)]. In the 3-dimensional case, verifying this explicitly matches Eq. (2.20).

Various determinant properties, such as Eq. (2.50), can be derived from Eq. (2.60). For instance, from high school mathematics:

1. If a row of the matrix A is the sum of two rows, then for the matrices B and C obtained by splitting that row:

$$|A| = |B| + |C|, \tag{2.61}$$

   and the same applies to columns [5].
2. If a matrix B is formed by swapping two rows of the matrix A:

$$|B| = (-)^\sigma |A|, \tag{2.62}$$

and the same applies to columns [5].

According to the fundamental definition of Eq. (2.49), the condition $|A| = 0$ corresponds to the scenario where the "projected basis vectors" collapse. This occurs when some basis vectors degenerate into the same vector upon projection, resulting in a situation where the "number of linearly independent basis vectors" changes. This is referred to as **rank deficiency**.

2.3.8 Cofactor Expansion and the Inverse Matrix

The definition of the determinant is given as:

$$|A| = \sum_P (-)^P a_{P_1,1} \cdot a_{P_2,2} \cdots a_{P_m,m} \cdots a_{P_N,N}.$$

By focusing on a specific term $a_{P(m),m}$, we can see that the determinant is linear with respect to the terms $a_{1m}, a_{2m}, \cdots$, and can be expressed as:

$$|A| = c_1^{(m)} \cdot a_{1m} + c_2^{(m)} \cdot a_{2m} + \cdots + c_N^{(m)} \cdot a_{Nm}. \tag{2.63}$$

Here, the j-th expansion coefficient $c_j^{(m)}$ can be extracted by setting all $\{a_{jk}\}_{k \neq m}$ to zero except for a_{jm},

$$a_{P_m,m} = \delta_{[P_m=j]}, \tag{2.64}$$

on the right-hand side. Substituting this, we obtain:

$$c_j^{(m)} = |A|_{\{a_{P_m,m}=\delta_{[P_m=j]}\}} \,. \tag{2.65}$$

Evaluating this expression, we find:

$$c_j^{(m)} = (-)^{(m+j)} |A_{jm}|, \tag{2.66}$$

as demonstrated in Sect. 9.2.2. Here, $|A_{jm}|$ represents the determinant of the matrix obtained by removing the m-th column and j-th row from A.

By substituting Eq. (2.66) into Eq. (2.63), we obtain:

$$\begin{aligned} |A| &= (-)^{1+m} a_{1m} \, |A_{1m}| + (-)^{2+m} a_{2m} \, |A_{2m}| + \cdots + (-)^{N+m} a_{Nm} \, |A_{Nm}| \\ &= \sum_l (-)^{l+m} a_{lm} \, |A_{lm}| . \end{aligned} \tag{2.67}$$

This is known as the **cofactor expansion** or the **Laplace expansion**.[28] For example, in the case of a 3×3 matrix, it gives the familiar determinant computation formula taught in undergraduate courses:

$$\begin{vmatrix} a_{11} & a_{12} & a_{13} \\ a_{21} & a_{22} & a_{23} \\ a_{31} & a_{32} & a_{33} \end{vmatrix} = a_{11} \begin{vmatrix} a_{22} & a_{23} \\ a_{32} & a_{33} \end{vmatrix} - a_{21} \begin{vmatrix} a_{12} & a_{13} \\ a_{32} & a_{33} \end{vmatrix} + a_{31} \begin{vmatrix} a_{12} & a_{13} \\ a_{22} & a_{23} \end{vmatrix} . \tag{2.68}$$

The formula expressed mathematically here as Eq. (2.67) corresponds to this computational method.[29]

In Eq. (2.67), note again that A_{jm} is the matrix obtained by removing the j-th row and m-th column from A. While the left-hand side is a scalar, the right-hand side involves index pairs (l, m), contracting the indices of a_{lm} and A_{lm}. From this appearance of indices, the determinant can also be written as:[30]

$$|A| = \sum_l (-1)^{l+m} \, a_{lm} \, |A_{lm}| = \sum_l B_{ml} \cdot a_{lm} \quad (\text{for fixed } m) \,. \tag{2.69}$$

[28] Pierre-Simon Laplace (1749.03–1827.03).

[29] Many modern undergraduate textbooks introduce determinants through concrete computational procedure such as Eq. (2.68) [1].

[30] Note the explicit annotation of "m-fixed". Misinterpreting this as a summation based on Einstein's summation convention would lead to incorrect handling.

Here, the definition

$$B_{ml} = (-1)^{l+m} |A_{lm}|$$

is motivated by the desire to perform contraction between a_{lm} and B over the inner index l. Expressing the indices as ij, we have:

$$B_{ij} = [\mathrm{Cof}\,(A)]_{ij} = (-1)^{i+j} \left|A_{ji}\right|,$$

where B_{ij} corresponds to A_{ji}, and the indices are transposed. The matrix $[\mathrm{Cof}\,(A)]_{ij}$ defined in this way is called the **cofactor matrix** of A. Again, the cofactor matrix A_{ji} refers to the matrix obtained by removing the j-th row and i-th column from A.

Using Eq. (2.69), we write:

$$|A| = B_{ml}a_{lm} \quad (\text{for fixed } m \quad , \quad \text{contracted } l)\,.$$

Let us consider the matrix formed by $B_{jl}a_{lm}$. By paying attention to the contraction over l, this corresponds to the jm component of the "matrix product of $B = \{B_{ij}\}$ and $A = \{a_{ij}\}$", given by:

$$\begin{aligned} B \cdot A &= \left\{(B \cdot A)_{jm}\right\} = \left\{B_{jl}a_{lm}\right\} \\ &= \begin{pmatrix} B_{1l}a_{l1} & * & \cdots & * \\ * & B_{2l}a_{l2} & & \\ \vdots & & \ddots & \\ * & \cdots & & B_{Nl}a_{lN} \end{pmatrix} \\ &= \begin{pmatrix} |A| & * & \cdots & * \\ * & |A| & & \\ \vdots & & \ddots & \\ * & \cdots & & |A| \end{pmatrix}. \end{aligned} \tag{2.70}$$

The off-diagonal elements are shown to be zero as derived in Sect. 9.2.3), so:

$$B \cdot A = |A| \begin{pmatrix} 1 & 0 & \cdots & 0 \\ 0 & 1 & & 0 \\ \vdots & & \ddots & \\ 0 & \cdots & & 1 \end{pmatrix} = |A| \cdot I. \tag{2.71}$$

Thus, we have:

$$B \cdot A = I \cdot |A| \quad , \quad \therefore \quad \frac{B}{|A|} \cdot A = I,$$

namely, **inverse matrix** of A is given by:

$$A^{-1} = \frac{\mathrm{Cof}\,(A)}{|A|}. \tag{2.72}$$

2.3.9 Concept of Matrix Diagonalization

Why is it necessary to use the information of an $N \times N$ square matrix

$$\hat{A} \sim \begin{pmatrix} a_{11} & a_{12} & \cdots & a_{1N} \\ a_{21} & \ddots & & \\ \vdots & & & \\ a_{N1} & \cdots & & a_{NN} \end{pmatrix}$$

instead of just N parameters to describe the linear projection $\hat{A}$ in an N-dimensional space? The reason for this can be traced back to Eq. (2.38):

$$\hat{A} \cdot \mathbf{e}_j = \sum_k c_k \cdot \mathbf{e}_k,$$

which serves as the starting point of the issue. Since the basis vectors **mix** with "other basis components" during the projection, when expressing a vector in the basis $\{\mathbf{e}_k\}$ as:

$$\mathbf{x} = \sum_k x_k \cdot \mathbf{e}_k, \tag{2.73}$$

the following computation arises:

$$\begin{aligned} \hat{A} \cdot \mathbf{x} &= \sum_k x_k \cdot \hat{A} \cdot \mathbf{e}_k = \sum_k x_k \cdot \left(\sum_m a_{mk} \cdot \mathbf{e}_m \right) \\ &= \sum_{m,k} a_{mk} x_k \cdot \mathbf{e}_m, \end{aligned}$$

$$\left(\hat{A} \cdot \mathbf{x} \right)_{\{\mathbf{e}_l\}} = \left[\begin{pmatrix} a_{11} & a_{12} & \cdots & a_{1N} \\ a_{21} & \ddots & & \\ \vdots & & & \\ a_{N1} & \cdots & & a_{NN} \end{pmatrix} \begin{pmatrix} x_1 \\ x_2 \\ \vdots \\ x_N \end{pmatrix} \right]_{\{\mathbf{e}_l\}}, \tag{2.74}$$

resulting in a complex "vector-matrix multiplication" operation.

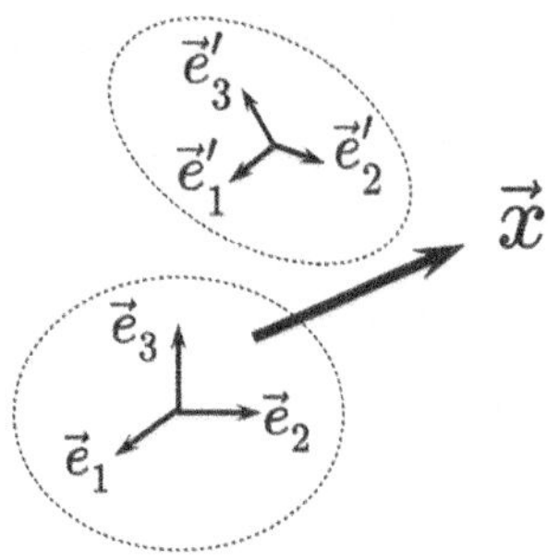

Fig. 2.6 Representing the same vector **x** in different coordinate systems, $\{\mathbf{e}_j\}$ and $\{\mathbf{e}_j'\}$

Now, let us express the same vector **x** in another basis:

$$\mathbf{x} = \sum_k x_k \cdot \mathbf{e}_k = \sum_k x_k' \cdot \mathbf{e}_k'.$$

(See Fig. 2.6).[31] If the new basis $\{\mathbf{e}_k'\}$ satisfies the property:

$$\hat{A} \cdot \mathbf{e}_j' = \lambda^{(j)} \mathbf{e}_j', \tag{2.75}$$

where it does **not mix** with others, then the mapping of the vector becomes:[32]

$$\hat{A} \cdot \mathbf{x} = \sum_k x_k' \cdot \hat{A} \cdot \mathbf{e}_k' = \sum_k x_k' \cdot \lambda^{(k)} \mathbf{e}_k',$$

$$i.e. \quad \left(\hat{A} \cdot \mathbf{x}\right)_{\{\mathbf{e}_l'\}} = \left[\begin{pmatrix} \lambda^{(1)} x_1' \\ \lambda^{(2)} x_2' \\ \vdots \\ \lambda^{(N)} x_N' \end{pmatrix}\right]_{\{\mathbf{e}_l'\}}.$$

This result is significantly simpler compared to Eq. (2.74).

This operation can be expressed as:

$$\left(\hat{A} \cdot \mathbf{x}\right)_{\{\mathbf{e}_l'\}} = \left[\begin{pmatrix} \lambda^{(1)} x_1' \\ \lambda^{(2)} x_2' \\ \vdots \\ \lambda^{(N)} x_N' \end{pmatrix}\right]_{\{\mathbf{e}_l'\}} = \left[\begin{pmatrix} \lambda^{(1)} & 0 & & \\ 0 & \lambda^{(2)} & & \\ & & \ddots & \\ & & & \lambda^{(N)} \end{pmatrix} \begin{pmatrix} x_1' \\ x_2' \\ \vdots \\ x_N' \end{pmatrix}\right]_{\{\mathbf{e}_l'\}}.$$

---

[31] Since some students may not fully grasp the concept that "the component representation of the same vector changes when expressed in a different basis", likely due to limited prior exposure, I have summarized the transformation of component representation using a 2D rotation matrix as an example in Sect. 9.2.4.

[32] The index j appearing in pairs on the right-hand side does not represent an inner product under Einstein's summation convention. To avoid confusion, we use the superscript (j).

Comparing this with Eq. (2.74), we see that:

$$\text{In the basis } \{\mathbf{e}_l\}, \begin{pmatrix} a_{11} & a_{12} & \cdots & a_{1N} \\ a_{21} & \ddots & & \\ \vdots & & & \\ a_{N1} & \cdots & & a_{NN} \end{pmatrix} \text{ represents the projection, whereas}$$

$$\text{in the basis } \left\{\mathbf{e}'_l\right\}, \begin{pmatrix} \lambda^{(1)} & 0 & & \\ 0 & \lambda^{(2)} & & \\ & & \ddots & \\ & & & \lambda^{(N)} \end{pmatrix} \text{ provides a diagonal representation.}$$

This suggests that the new coordinate satisfying Eq. (2.75) simplifies the representation of the projection. For detailed derivations, refer to Sect. 9.2.5. In summary, solving Eq. (2.75)[33] gives N **eigenvectors**:

$$\mathbf{e}'_1 = \begin{pmatrix} p_{11} \\ p_{21} \\ \vdots \\ p_{N1} \end{pmatrix}_{\{\mathbf{e}_j\}}, \quad \mathbf{e}'_2 = \begin{pmatrix} p_{12} \\ p_{22} \\ \vdots \\ p_{N2} \end{pmatrix}_{\{\mathbf{e}_j\}}, \quad \cdots, \quad \mathbf{e}'_N = \begin{pmatrix} p_{1N} \\ p_{2N} \\ \vdots \\ p_{NN} \end{pmatrix}_{\{\mathbf{e}_j\}},$$

which, when arranged as columns, form the matrix:

$$P = \left(\mathbf{e}'_1, \mathbf{e}'_2, \cdots, \mathbf{e}'_N\right) = \begin{pmatrix} p_{11} & p_{12} & \cdots & p_{1N} \\ p_{21} & \ddots & & \\ \vdots & & & \\ p_{N1} & \cdots & & p_{NN} \end{pmatrix}.$$

Using this square matrix P, the representing matrix $\hat{A}$ in the "old coordinate system" A and in that in the "new coordinate system" A' are related by:

$$A' = P^{-1}AP.$$

[33] The equation is called **eigenvalue equation**.

Evaluating this, we find that:

$$A' = \begin{pmatrix} \lambda^{(1)} & 0 & & \\ 0 & \lambda^{(2)} & & \\ & & \ddots & \\ & & & \lambda^{(N)} \end{pmatrix},$$

which is a "**diagonal matrix** with eigenvalues as its elements".

Thus, we aim to find new basis sets such that

$$\hat{A} \cdot \mathbf{e}'_j = \lambda^{(j)} \mathbf{e}'_j \tag{2.76}$$

and expand a general vector in the form

$$\mathbf{x} = \sum_k x'_k \cdot \mathbf{e}'_k .$$

This provides a compact representation for the projection.

The quantities $\left\{\lambda^{(j)}\right\}$ satisfying Eq. (2.76) are called **eigenvalues**, while $\left\{\mathbf{e}'_j\right\}$ are called **eigenvectors**. The problem of finding eigenvalues and eigenvectors is referred to as the **eigenvalue problem**. It can be shown that this problem reduces to solving the equation

$$\det (A - \lambda I) = 0, \tag{2.77}$$

known as the **characteristic equation** (see Sect. 9.2.6). Since this involves the determinant of an $N \times N$ matrix, it results in an N-th degree equation in λ. Thus, we obtain N solutions: $\left\{\lambda^{(1)}, \lambda^{(2)}, \cdots, \lambda^{(N)}\right\}$. For each $\left\{\lambda^{(j)}\right\}$, there exist N corresponding eigenvectors $\left\{\mathbf{e}'_j\right\}$ satisfying Eq. (2.76).

The determinant was introduced as a measure of how the volume of an N-dimensional parallelepiped changes when projected by the mapping $\hat{A}$ (Sect. 2.3.5). From the perspective using the basis of eigenvectors $\left\{\mathbf{e}'_j\right\}$, the mapping $\hat{A}$ scales each axis by $\lambda^{(j)}$. Hence,

$$(\det A) = \prod_{j=1}^{N} \lambda^{(j)} \tag{2.78}$$

is anticipated. This result can also be derived (see Sect. 9.2.5).

Representation Theory of Groups
The essence of group representation theory can be described as an "analogy to vector spaces". Spatial symmetry operations can be represented as projection operators that map elements, such as vertices, onto one another. Since the elements form a group and transform into each other, this structure resembles that of vector spaces, where projections remain confined within the same "village". As a result, the space representing the group is expected to be expressible as a linear combination of some "basis set". Consequently, the subsequent formulation parallels that of linear mappings: constructing "projection representation matrices" based on appropriately defined bases. This naturally leads to the idea of finding a new diagonalized basis to disentangle and simplify the relationships between bases (reduction operations in group theory) [1]. The eigenvectors obtained in this process (untangled and clearly defined diagonal bases) correspond to the irreducible representations of the group.

2.4 Differential Equations and Waves

In this section, we introduce fundamental concepts related to differential equations. The inclusion of this topic serves the purpose of reinforcing the understanding of wave equations and the representation of waves. Waves are a concept that tends not to be well-established among students from fields like information science, particularly those who may lack a solid foundation in general undergraduate mathematics [1]. As a prelude to the wave equation discussed later, we draw attention to the idea of "describing governing laws as relationships between differential quantities". Additionally, the chain rule for partial derivatives and the treatment via basis function expansions frequently appear in Sect. 4 and beyond, so these concepts are introduced here.

2.4.1 Differential Equations: Estimating Solutions

When one thinks of equations, the first thing that comes to mind is the familiar "equation for values" from secondary school. For example, solving the equation $3x = 6$ yields $x = 2$ as the **value** that satisfies the equality. On the other hand, a "functional equation" that seeks a **function** satisfying the equality takes the form of a differential equation such as:

$$\frac{df(x)}{dx} = 3f(x) . \tag{2.79}$$

This equation can be solved as follows: Eq. (2.79) can be interpreted as "find a function that is proportional to its derivative". Normally, the original function and its derivative "live in different world". For example, differentiating x^3 yields $3x^2$, and differentiating x^2 yields $2x$. If x is interpreted as length, x^3 resides in the world of volume, while x^2 resides in the world of area. Thus, Eq. (2.79) asks for a function in special situation that "remains in the same world even after differentiation".

Recalling the differentiation rule for exponential functions,

$$\frac{d}{dx}\left[e^{ax}\right] = a \cdot e^{ax},$$

we see that exponential functions possess the unique property of "remaining in the same world even after differentiation". Substituting an exponential function of the form

$$f(x) = A \cdot e^{ax} \tag{2.80}$$

as a promising candidate solution into Eq. (2.79), we find:

$$\frac{df(x)}{dx} = a \cdot A \cdot e^{ax} = a \cdot f(x) \overset{!}{=} 3 \cdot f(x) \quad , \quad \therefore \quad a = 3.$$

Hence, the solution is found as

$$f(x) = A \cdot e^{3x}.$$

Using a similar approach, the differential equation

$$\frac{d^2 f(x)}{dx^2} + 3\frac{df(x)}{dx} + 2f(x) = 0 \tag{2.81}$$

can also be solved. The condition that "a combination of the second derivative, the first derivative, and the original function sums to zero" holds only when all these terms "live in the same world". Assuming a solution of the form, Eq. (2.80), and substituting, we obtain:

$$a^2 + 3a + 2 = 0 \quad , \quad \therefore \quad (a+2)(a+1) = 0.$$

The possible solutions are $a = -1$ and $a = -2$. Denoting the solutions by $f_1(x)$ and $f_2(x)$, both satisfy Eq. (2.81). Therefore,

$$f(x) = f_1(x) + f_2(x)$$

is also a solution to the differential equation, being called **linearity**. With $f_1(x) = A \cdot e^{-x}$ and $f_2(x) = B \cdot e^{-2x}$, the general solution can be written as

$$f(x) = A \cdot e^{-x} + B \cdot e^{-2x}.$$

While the explanation above may make solving differential equations seem simple, for equations like

$$\frac{d^2 f(x)}{dx^2} + 3\frac{df(x)}{dx} + 2f^2(x) = 0,$$

the situation becomes far more complex. Here, the discussion shifts to "the squared term of the sought function and its derivatives living in the same world", making the approach used in Eq. (2.80) unsuitable. While it is manageable when the functional form of the solution can be easily guessed, in general, this is quite challenging. Analytical solutions are available only for a very limited class of differential equations.[34]

2.4.2 Describing Phenomena Using Differential Equations

Imagine two students each throwing a ball. The trajectories of the balls differ depending on who threw them. From the perspective of modern science, we understand that "while the individual trajectories differ, they all obey the same governing laws". However, to someone without the foundational knowledge of modern science, such as a primitive observer, it would not be unreasonable to think, "the trajectories are different, so these must be different phenomena". Even if they suspect otherwise, articulating a proper counterargument would be a difficult task. This leads us to consider how to describe and comprehend the dichotomy between "distinct individual events" and the "unifying principles governing them".

In the context of mechanics, the governing law is Newton's equation, and the fact that "the trajectories differ for individual students" stems from differences in the initial conditions[35] provided to the equation. Newton's equation, which is the governing equation, is expressed as a second-order differential equation. Therefore, when solving the equation for a specific event, twice integrations are required, introducing two integration constants. These constants correspond to the "initial position and initial velocity for each individual event", and their differences

[34] For linear differential equations with constant coefficients, the method described in the text, which assumes an exponential form for the solution, is applicable. When the coefficients involve powers of x, solutions are often sought as series expansions, leading naturally to the introduction of special functions [1].

[35] For example, differences in initial velocity, starting position, or throwing angle.

account for the distinct trajectories observed in each case. This understanding can be formalized as follows:

- Differences between individual cases $\rightarrow$ Differences in integration constants,
- Common governing law $\rightarrow$ Descriptions based solely on relationships between differential quantities.

Through the operation of differentiation, the **integration constants**, which are features of individual events, are effectively **washed away**, leaving only the essential common laws described. This approach is expressed with the nuance of the term "describing dynamics".

2.4.3 Boundary Value and Eigenvalue Problems

As an example, let us consider the differential equation

$$\frac{d^2 f}{dx^2} + \lambda^{(\text{given})} \cdot f = 0, \tag{2.82}$$

where the constant $\lambda^{(\text{given})}$ indicates that λ is a *given* constant. By assuming a solution similar to Eq. (2.80) and obtaining $a^2 + \lambda = 0$, we solve this as

$$f(x) = A \cdot e^{i\sqrt{\lambda}\cdot x} + B \cdot e^{-i\sqrt{\lambda}\cdot x}. \tag{2.83}$$

Now, consider a similar setup with the equation

$$\frac{d^2 f}{dx^2} + \lambda^{(\text{unknown})} \cdot f = 0, \tag{2.84}$$

and the **boundary conditions**

$$f(\pi) = f(0) = 0. \tag{2.85}$$

The goal is to find the function $f(x)$ and the constant $\lambda^{(\text{unknown})}$ that satisfy these conditions. Here, $\lambda^{(\text{unknown})}$ indicates an *unknown* value to be determined under the given conditions, distinguishing it from the *given* $\lambda^{(\text{given})}$.

Since the equation itself is identical to Eq. (2.82), the solution Eq. (2.83) satisfies Eq. (2.84). Substituting $x = 0$ and applying the boundary condition $f(0) = 0$ gives the relationship $A + B = 0$, resulting in

$$f(x) = A \cdot \left[e^{i\sqrt{\lambda}\cdot x} - e^{-i\sqrt{\lambda}\cdot x}\right] = C \cdot \sin\left[\sqrt{\lambda} \cdot x\right],$$

where C is a complex coefficient. Applying the boundary condition $f(\pi) = 0$ leads to

$$C \cdot \sin\left[\sqrt{\lambda} \cdot \pi\right] \stackrel{!}{=} 0,$$

$$\therefore \quad \sqrt{\lambda} \cdot \pi \stackrel{!}{=} k \cdot \pi \quad , \quad k = 0, 1, 2, \cdots ,$$

$$\therefore \quad \lambda = k^2 = 0, 1, 4, 9, \cdots .$$

Thus, the possible values of the unknown $\lambda^{(\text{unknown})}$ are obtained.

Revisiting the boundary value problem:

$$\frac{d^2 f}{dx^2} + \lambda^{(\text{unknown})} \cdot f = 0,$$

its solutions are obtained as

$$\lambda^{(\text{unknown})} = k^2 \quad , \quad f(x) = C \cdot \sin(kx) =: f_k(x) \quad , \quad (k = 0, 1, 2, \cdots) .$$

The subscript k has been appended to f_k, emphasizing the indexed pair $\{\lambda(k), f_k(x)\}_{k=1,2,\ldots}$, representing **several possible modes**. It is crucial to note that the setting of the boundary conditions determines and restricts the range of possible eigenvalues.

Equation (2.84) can be rewritten as

$$\frac{d^2}{dx^2} f(x) = \mu \cdot f(x), \tag{2.86}$$

by defining $\mu = -\lambda$. Here, the operator $(d^2/dx^2) = \hat{A}$ satisfies

$$\frac{d^2}{dx^2}\left[f_1(x) + f_2(x)\right] = \frac{d^2}{dx^2} f_1(x) + \frac{d^2}{dx^2} f_2(x),$$

indicating that it is a linear operator. Writing Eq. (2.86) as

$$\hat{A} \cdot f(x) = \mu \cdot f(x),$$

implies that "the result of applying a linear operator to a function is proportional to the function itself". This is analogous to Eq. (2.76). Functions f satisfying this equation are called **eigenfunctions**, and λ are called **eigenvalues**. Such problems are referred to as **eigenvalue problems for differential equations**.

In Sect. 2.3, we discussed how abstract linear operators like $\hat{A}$ can be represented in matrix form. Here, we explain how the "eigenvalue problem for differential equations" can be translated into the "eigenvalue problem for matrices" through such representations.

2.4.4 Fourier Series and Basis Set Expansion

Consider a periodic function $f(x)$ with a period of 2π. It seems plausible to express it in terms of $\sin x$ and $\cos x$:

$$f(x) \sim a \cdot \cos(x) + b \cdot \sin(x).$$

However, since higher harmonics such as $\sin(kx)$, where k is an integer, can also be considered as functions with a period of 2π, the idea extends to include them:

$$f(x) = \sum_{k=1}^{\infty} [a_k \cdot \cos(kx) + b_k \cdot \sin(kx)]. \tag{2.87}$$

This approach is called the **Fourier series.**[36]

Using Euler's formula as mentioned in Eq. (2.7), substituting $-x$ for x, and adding or subtracting e^{ix} and e^{-ix}, we obtain:

$$\cos x = \frac{1}{2} \cdot e^{ix} + \frac{1}{2} \cdot e^{-ix}, \quad \sin x = -\frac{i}{2} \cdot e^{ix} + \frac{i}{2} \cdot e^{-ix}.$$

This shows that $\sin x$ and $\cos x$ can be considered as derived from the more fundamental components e^{ix} and e^{-ix}. Based on this perspective, Eq. (2.87) can be rewritten as:

$$f(x) = \sum_{k=1}^{\infty} \left[\alpha_k \cdot e^{ikx} + \beta_k \cdot e^{-ikx}\right],$$

which represents $f(x)$ in terms of the fundamental components e^{ikx} and e^{-ilx}. By extending the summation over k to include negative values, this can further be expressed as:

$$f(x) = \sum_{k=-\infty}^{\infty} c_k \cdot e^{ikx}. \tag{2.88}$$

Comparing this equation to the vector basis expansion:

$$\mathbf{x} = \sum_{k} c_k \cdot \mathbf{e}_k,$$

[36] Jean Baptiste Joseph Fourier (1768.03–1830.05). The idea itself was known before Fourier, through Euler and Lagrange. Fourier's contribution was to assert that this framework applies to "arbitrary functions". This claim was met with skepticism and criticism, particularly from Lagrange and other French mathematicians at the time [1].

we notice that they share the same structure. Thus, the functions e^{ikx} in this series expansion are referred to as **basis set functions**.

By acting the "conjugate of the basis function" e^{-imx} on both sides of Eq. (2.88), with a factor of $(1/L)$ and integrating over a periodic interval L, we have:[37]

$$\frac{1}{L}\int_L dx \cdot e^{-imx} f(x) = \frac{1}{L}\int_L dx \cdot e^{-imx}\left(\sum_k c_k \cdot e^{ikx}\right)$$
$$= \sum_k c_k \cdot \frac{1}{L}\int_L dx \cdot e^{i(k-m)x}. \tag{2.89}$$

The integral appearing in the above equation evaluates to:

$$\frac{1}{L}\int_L dx \cdot e^{i(k-m)x} = \delta_{mk}. \tag{2.90}$$

This is because, for $m \neq k$, higher harmonics of e^{ix} remain, and integrating a trigonometric function over a period yields zero. For $m = k$, $e^{i(m-k)x} = 1$, so the integral value is L, which cancels with the denominator L, resulting in 1.[38] Substituting Eq. (2.90) into Eq. (2.89) gives:

$$c_m = \frac{1}{L}\int_L dx \cdot e^{-imx} f(x). \tag{2.91}$$

The above is a summary of Fourier series expansion. Here, let us now introduce the notation:

$$\langle g|f\rangle := \int_L dx \cdot g^*(x) f(x),$$

where multiplying $f(x)$ by the conjugate of $g(x)$ and integrating over the interval is denoted as "taking the inner product with $g(x)$", and the result is written as $\langle g|f\rangle$. Adjusting the Fourier series basis functions e^{imx} with the normalization factor of $1/\sqrt{L}$, we define:

$$\frac{1}{\sqrt{L}} \cdot e^{imx} \sim |m\rangle, \quad \frac{1}{\sqrt{L}} \cdot e^{-imx} \sim \langle m|.$$

[37] The index k is used as a dummy variable in substituting $f(x)$, so the conjugate uses e^{-imx} with the index m. At this stage, L can be assumed to be 2π. The factor $(1/L)$ is included for normalization, as explained later.

[38] The factor $(1/L)$ was included to ensure proper normalization.

Here, $|m\rangle$ denotes the basis function, and $\langle m|$ its conjugate. Using this notation, the series expansion in Eq. (2.88) becomes:

$$f(x) = \sum_k c_k \cdot \frac{e^{ikx}}{\sqrt{L}}$$

$$\rightarrow \quad |f\rangle = \sum_k c_k \cdot |k\rangle .$$

This representation is known as the **bra-ket notation**.[39] With bra-ket notation, Eqs. (2.90) and (2.91) can be rewritten as:

$$\langle m|k\rangle = \delta_{mk}, \quad c_m = \langle m|f\rangle .$$

By expressing spatial vectors using the notation $\mathbf{x} = |x\rangle$ and their inner product as $\mathbf{a} \cdot \mathbf{b} = \langle a|b\rangle$, we can write the properties of spatial vectors as:

$$|x\rangle = \sum_k c_k \cdot |e_k\rangle, \quad \langle e_m|e_k\rangle = \delta_{mk}, \quad x_m = \langle e_m|x\rangle ,$$

as described in Sect. 2.3.1. Comparing this to the Fourier series expansion:

$$|f\rangle = \sum_k c_k \cdot |k\rangle, \quad \langle m|k\rangle = \delta_{mk}, \quad c_m = \langle m|f\rangle , \tag{2.92}$$

we observe a striking similarity between the two forms.

The essence of why the Fourier series expansion resembles the structure of spatial vectors lies in the generalizability of this format beyond Fourier expansions: The ability to extract components as $c_m = \langle m|f\rangle$, by projecting onto a basis $|m\rangle$, is rooted in the **orthogonality of the basis**, $\langle m|k\rangle = \delta_{mk}$. This orthogonality, in turn, is defined by the *inner product* over a specified domain L. In the case of Fourier expansions, the domain L corresponds to the fundamental period of the periodic function. The basis functions e^{imx}, being trigonometric functions (Sect. 2.1.3), have the property that their integral over the fundamental period vanishes unless they coincide, establishing their orthogonality.

Depending on the problem, the integration domain Ω can vary, such as a circle, sphere, or intervals like $x = [-1, 1]$ or $x = [0, \infty)$. The inner product is generally defined as:

$$\langle g|f\rangle := \int_\Omega dx \cdot w(x) \cdot g^*(x) f(x),$$

[39] Further details on bra-ket notation are provided in Sect. 9.2.7, but for now, this level of rough introduction suffices.

where $w(x)$ is a weight function introduced based on the problem's requirements.[40] With this inner product definition, if functions $\{|k\rangle\}$ satisfy the orthogonality condition $\langle m|k\rangle = \delta_{mk}$, Eq. (2.92) holds with the same reasoning.

The Fourier series expansion:

$$f(x) = \sum_k c_k \cdot \frac{e^{ikx}}{\sqrt{L}},$$

is a specific case of the more general basis set expansion:

$$f(x) = \sum_k c_k \cdot \chi_k(x), \tag{2.93}$$

where the basis functions $\{\chi_k(x)\}$ satisfy orthogonality:

$$\langle \chi_k|\chi_l\rangle := \int_\Omega dx \cdot w(x) \cdot \chi_k^*(x)\chi_l(x) = \delta_{kl}.$$

The Fourier series is an example of an expansion using an orthogonal plane wave basis set. Other examples include expansions with Bessel functions, Legendre polynomials, Laguerre polynomials, and Chebyshev polynomials[1].

2.4.5 Eigenvalue Problems and Basis Function Expansions

With the preparations above, we can now understand how the *eigenvalue problem of the differential equation* given by Eq. (2.86):

$$\frac{d^2}{dx^2} f(x) = \mu \cdot f(x) \tag{2.94}$$

can be reduced to the *eigenvalue problem for matrices*.

By applying the basis function expansion in Eqs. (2.93)–(2.94) and substituting, we get:

$$\sum_l c_l \cdot \frac{d^2}{dx^2}\chi_l(x) = \mu \sum_l c_l \cdot \chi_l(x).$$

40 For instance, setting $w(x) = x^2$ expresses the situation where regions with larger x values contribute more significantly to the integral.

Following the method used in Eq. (2.89), we multiply both sides by the conjugate $\chi_k^*(x)$ and integrate over the domain:

$$\int_\Omega dx \cdot \left[\sum_l c_l \chi_k^*(x) \frac{d^2}{dx^2}\chi_l(x)\right] = \int_\Omega dx \cdot \left[\mu \sum_l c_l \chi_k^*(x)\chi_l(x)\right].$$

Using the inner product notation:

$$\int_\Omega dx \cdot \chi_k^*(x) f(x) = \langle \chi_k | f \rangle ,$$

this can be rewritten as:

$$\sum_l c_l \cdot \left\langle \chi_k \middle| \frac{d^2}{dx^2}\chi_l \right\rangle = \mu \sum_l c_l \langle \chi_k | \chi_l \rangle .$$

Since the second derivative operator is linear, it can be written as $\hat{A}$. Using the orthogonality of the basis functions, $\langle \chi_k | \chi_l \rangle = \delta_{kl}$, we get:

$$\sum_l c_l \cdot \langle \chi_k | \hat{A}\chi_l \rangle = \mu \sum_l c_l \cdot \delta_{kl} = \mu \cdot c_k. \tag{2.95}$$

$|\hat{A}\chi_l\rangle = \left|\frac{d^2}{dx^2}\chi_l\right\rangle$ represents the function obtained by taking the second derivative of the l-th basis function. Since this resulting function also resides in the same space, it can once again be expanded in terms of the basis functions as:[41]

$$\left|\hat{A}\chi_l\right\rangle = \left|\frac{d^2}{dx^2}\chi_l\right\rangle = \sum_m a_{ml} |\chi_m\rangle .$$

Using the orthogonality of the basis functions, the coefficients are given by:

$$\langle \chi_k | \hat{A}\chi_l \rangle = \sum_m a_{ml} \langle \chi_k | \chi_m \rangle = \sum_m a_{ml} \cdot \delta_{km} = a_{kl} .$$

Substituting this result into Eq. (2.95), we obtain:

$$\sum_l a_{kl} \cdot c_l = \mu \cdot c_k .$$

[41] Make sure that by now, you can independently and correctly assign subscripts like a_{kl} without being told. It is a_{ml}, not a_{lm}. Recall the comment made at the end of Sect. 2.3.3 for clarification.

If one is familiar with the subscript rules for matrix-vector multiplication, it becomes evident that this corresponds to the *eigenvalue problem for matrices*,

$$A \cdot \mathbf{c} = \mu \cdot \mathbf{c}, \tag{2.96}$$

where $A = \{a_{kl}\}$ represents a matrix with elements a_{kl}, and $\mathbf{c} = \{c_j\}$ is a vector.

If we adopt the notation $\langle \chi_i | \hat{A} | \chi_j \rangle$ for $\langle \chi_i | \hat{A} \chi_j \rangle$, the expression becomes easier to remember as $a_{ij} = \langle \chi_i | \hat{A} | \chi_j \rangle$. Anyway, it is important to note that this is actually:

$$a_{ij} = \int_\Omega dx \cdot \left[\chi_i^*(x) \frac{d^2}{dx^2} \chi_j(x) \right],$$

a numerical value obtained by performing the integral. Once the matrix elements $A = \{a_{ij}\}$ are prepared as values, we can represent the function $f(x)$ and its coefficients as:

$$f(x) = \sum_l c_l \cdot \chi_l(x) \quad \leftrightarrow \quad \begin{pmatrix} c_1 \\ c_2 \\ \vdots \end{pmatrix} =: \mathbf{c},$$

identifying the function with the vector $\mathbf{c}$. In this framework, the eigenvalue problem for the differential equation, Eq. (2.86), is reduced to the eigenvalue problem for the matrix, Eq. (2.96).

Supercomputer Performance Rankings

Many scientific and technological simulation problems are formulated as differential equations. Supercomputers designed to solve such simulations are ranked twice a year in the **Top500**, a global performance ranking. The ranking is based on the speed at which supercomputers can execute linear algebra computations, such as matrix-vector multiplications, calculating inverse matrices, or finding matrix eigenvalues [6]. At first glance, it might seem pointless to rank performance using these "seemingly irrelevant and narrowly focused computational methods" that appear far removed from real-world simulation problems. However, as discussed in this section, many problems expressed as differential equations are reformulated and solved as matrix computations. Thus, **the performance of linear algebra computations** serves as a highly universal and meaningful performance metric.

2.4.6 Separation of Variables for Partial Differential Equations

Consider the boundary value problem for the partial differential equation:

$$\frac{\partial u(x,t)}{\partial t} = 2\frac{\partial u(x,t)}{\partial x} \quad , \quad u(x,0) = e^{3x}.$$

The solution $u(x,t)$ depends on both time and space, and in general, the effects of these variables may be intricately intertwined. However, if we assume the functional form:

$$u(x,t) = F(x)\,G(t),$$

where the solution separates into a product of spatial and temporal components, we can simplify the problem as follows: Substituting the assumed **separable form** into the equation yields:

$$F(x)\frac{\partial G(t)}{\partial t} = 2\frac{\partial F(x)}{\partial x}G(t),$$

$$\therefore \quad \frac{dG(t)}{dt}\frac{1}{G(t)} = 2\frac{dF(x)}{dx}\frac{1}{F(x)}.$$

Here, since the left-hand side depends only on t, and the right-hand side only on x, the equality can hold for all x and t only if both sides are equal to a constant, say λ. Thus, we have:

$$\frac{dG(t)}{dt}\frac{1}{G(t)} = 2\frac{dF(x)}{dx}\frac{1}{F(x)} = \lambda.$$

This gives the following ordinary differential equations for $F(x)$ and $G(t)$:

$$\frac{dF(x)}{dx} = \frac{\lambda}{2}\cdot F(x), \quad \frac{dG(t)}{dt} = \lambda \cdot G(t).$$

The problem has now been **separated into variables**, reducing it to two independent differential equations. These equations can be solved as described in Sect. 2.4.1:

$$F(x) = A\cdot\exp\left[\frac{\lambda}{2}\cdot x\right], \quad G(t) = B\cdot\exp[\lambda\cdot t].$$

Combining these results to construct $u(x,t)$:

$$u(x,t) = F(x)\,G(t) = AB\cdot\exp\left[\lambda\left(\frac{x}{2}+t\right)\right].$$

From the boundary condition:

$$u(x,0) = AB \cdot \exp\left[\lambda\left(\frac{x}{2}\right)\right] = e^{3x},$$

we determine the constants:

$$AB = 1, \quad \lambda = 6.$$

Finally, the solution becomes:

$$u(x,t) = \exp\left[6\left(\frac{x}{2} + t\right)\right].$$

This procedure of separating variables in a partial differential equation to reduce it to a set of ordinary differential equations will frequently appear in subsequent discussions.

2.4.7 Description of Variations/Partial Derivatives and Chain Rule

When a variable x depends on time, it can be expressed as $x(t)$, with its rate of change given by $dx(t)/dt$. Now, consider a tank with water flowing in and out through two hoses, where the flow rates of each hose are f_1 and f_2, and the water level in the tank is h. Since the water level depends on the flow rates, it can be expressed as $h(f_1, f_2)$. If the flow rates themselves vary with time, i.e., $f_1(t)$ and $f_2(t)$, then the time dependence of $h(f_1(t), f_2(t))$ is expressed as follows:

$$h(f_1(t), f_2(t)) = h(t).$$

This describes the "overall time dependence of the water level in the tank" (Fig. 2.7).

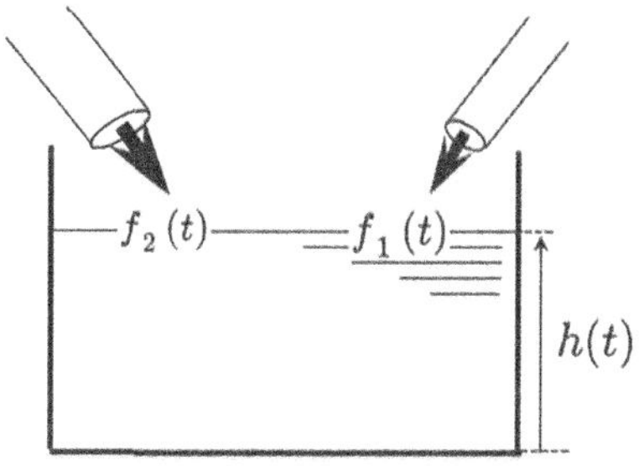

Fig. 2.7 Changes in water level $h(t)$. Since the water level depends on the flow rates from each hose, $f_1(t)$ and $f_2(t)$, the water level can be expressed as $h(f_1(t), f_2(t)) = h(t)$

Let us now consider how the rate of change of the water level, $dh(t)/dt$, is determined:

$$\left[\text{Total variation;}\frac{dh(t)}{dt}\right] = [\text{Variation via } f_1(t)] + [\text{Variation via } f_2(t)]\,.$$

For the variation via $f_1(t)$, it can be expressed as:

$$\begin{aligned}&(f_1(t)\text{ -induced variation})\\&\quad= [\text{How affecting the variations in } f_1 \text{ to } h(f_1, f_2)]\\&\quad\quad\times[\text{Extent of variation in } f_1 \text{ itself}]\\&\quad= \frac{\partial h}{\partial f_1} \times \frac{df_1}{dt}.\end{aligned}$$

Similarly, for f_2, we obtain:

$$\frac{dh\,(t)}{dt} = \frac{\partial h}{\partial f_1} \cdot \frac{df_1}{dt} + \frac{\partial h}{\partial f_2} \cdot \frac{df_2}{dt}.$$

This rule aligns precisely with the intuitive notion that "∂f_1" in the numerator and denominator cancel each other out, making it a "symbolically natural" notation. This rule, which appears as if the terms in the numerator and denominator cancel out, is called the **chain rule**.

Leibniz Notation
The differential notation dy/dx, originally introduced to represent "rates of change", is not to be treated as a simple division where cancellation of numerator and denominator is inherently valid. However, Leibniz notation is an elegant symbolic representation where such cancellation often aligns with intuitive expectations (e.g., $dz/dt = (dz/dy)/(dt/dy)$). Balancing "semantic correctness" and "symbolic intuition", as discussed in Sect. 2.2.4, is crucial. In mathematics, constructing symbolic systems where semantically provable relationships align with symbolically natural expectations is considered a hallmark of beauty. The validity of the "cancellation" in Leibniz notation can often be traced back to the definition of derivatives as "limits of quotients", enabling such symbolic manipulations.

When f_1, f_2, and more variables are involved, a similar argument applies to a general dependence $h(t) = h\left(f_1(t), \cdots, f_N(t)\right)$:

$$\frac{dh(t)}{dt} = \frac{\partial h}{\partial f_1}\frac{df_1}{dt} + \cdots + \frac{\partial h}{\partial f_j}\frac{df_j}{dt} + \cdots + \frac{\partial h}{\partial f_N}\frac{df_N}{dt} = \sum_l \frac{\partial h}{\partial f_l} \cdot \frac{df_l}{dt}.$$

Reversing the left and right-hand sides:

$$\sum_l \frac{\partial h}{\partial f_l} \cdot \frac{df_l}{dt} = \frac{\partial h}{\partial f_l} \cdot \frac{df_l}{dt} = \frac{dh}{dt}. \tag{2.97}$$

The left-hand side symbolically appears as a "common denominator", yielding the right-hand side.[42] This relationship, frequently used later, is important to familiarize oneself with.

Finally, consider the case where the tank connected to the hoses is elastic, like a rubber balloon, and its own expansion or contraction affects the water level over time. In this case, the time variation of the water level incorporates the effect of the tank's changes, represented as:

$$h(t) = h\left(t, f_1(t), f_2(t)\right) .$$

Here, the first t in the arguments $(t, f_1(t), f_2(t))$ represents the **explicit time dependence** of the tank itself, while $(f_1(t), f_2(t))$ represents the **implicit dependence** via the hoses. The rate of change of h is hence expressed as:

$$\frac{dh(t)}{dt} = \frac{\partial h}{\partial t} + \frac{\partial h}{\partial f_1} \cdot \frac{\partial f_1}{\partial t} + \frac{\partial h}{\partial f_2} \cdot \frac{\partial f_2}{\partial t} .$$

Note the distinction between the total derivative $dh(t)/dt$ on the left-hand side and the partial derivative $\partial h/\partial t$ on the right-hand side. The partial derivative accounts only for the contribution from explicit dependence, while the total derivative includes contributions from all channels, both explicit and implicit.

2.4.8 *Wave Equation*

A wave can be described as "a waveform $f_1(x)$ propagating while maintaining its shape". Let us consider how to express this mathematically. Assume a waveform described at time $t = 0$ by $y = f_1(x)$, propagates along the x-axis at a velocity v in the positive direction.

---

[42] The middle equation employs Einstein's summation convention.

Recall that the expression $y = f(x)$ can be interpreted as: "When focusing on a specific point x, the y-value observed there is...". This interpretation corresponds to *fixed-point observation*. With this in mind, as the waveform moves to the right over a time t, it means: "The y-value that was located vt to the left of x at $t = 0$ is now observed at position x at time t". Therefore, shifting the argument of the waveform $f_1(x)$ in the x-direction as $x \to (x - v \cdot t)$ and expressing it as $f_1(x - v \cdot t)$ mathematically represents "a waveform propagating to the right over time t". The above describes a "wave propagating to the right", but waves generally include those propagating to the left as well. In general, a wave is expressed as a function whose time evolution is represented by:

$$\xi(x, t) = f_1(x - v \cdot t) + f_2(x + v \cdot t), \tag{2.98}$$

which includes components propagating both to the right and to the left.

As discussed in Sect. 2.4.2, the goal is not to describe waves at the level of individual functions but to formulate them as differential equations that "wash away" specific constants of integration tied to individual phenomena. The differential equation whose solution reduces to Eq. (2.98) is referred to as the **wave equation**. This equation describes functions that take (x, t) as arguments, leading to a form that combines time derivatives and spatial derivatives. To first show the answer to what kind of equation it is:

$$\frac{\partial^2 \xi(x, t)}{\partial t^2} = v^2 \cdot \frac{\partial^2 \xi(x, t)}{\partial x^2}. \tag{2.99}$$

This is the **wave equation**, which describes the dynamics of waves. Quite a few textbooks present Eq. (2.99) without much context, stating, "Eq. (2.98) indeed satisfies Eq. (2.99), so Eq. (2.99) is the wave equation", relying on a highly heuristic explanation. However, such an introduction feels somewhat unsatisfactory. In this book, we aim to explain the wave equation through a more systematic and non-heuristic discussion.

A distinctive feature of the mathematical description of waves is that the two-variable function

$$\xi(x, t) = f_1(x - v \cdot t) + f_2(x + v \cdot t)$$

essentially takes the form where "the two variables are encapsulated within the arguments of the single-variable functions $f_j(x)$". The objective is to describe this "feature" in terms of differential relationships. The approach to extracting this feature is based on the following perspective: Both the time partial derivative and the spatial partial derivative of ξ, if taken to sufficiently high orders, will inevitably encounter the single-variable structure:

$$\xi(x, t) = \xi(x \pm vt) = \xi(\eta),$$

where the shared structure of derivatives with respect to the single variable η will result in the equality where the time and the spatial partial derivatives are the linked by some coefficients. Evaluating the partial derivatives of ξ with respect to time and space yields:

$$\frac{\partial \xi}{\partial t} = \frac{\partial \xi}{\partial \eta}\frac{\partial \eta}{\partial t} = \pm v \frac{\partial \xi}{\partial \eta} \quad , \quad \frac{\partial \xi}{\partial x} = \frac{\partial \xi}{\partial \eta}\frac{\partial \eta}{\partial x} = \frac{\partial \xi}{\partial \eta}.$$

Through $\partial \xi / \partial \eta$, we obtain:

$$\left(\frac{\partial \xi}{\partial \eta} =\right) \quad \frac{\partial \xi}{\partial x} = \pm \frac{1}{v}\frac{\partial \xi}{\partial t}$$

which appears to be the desired relationship where the time and spatial partial derivatives "meet". However, the $\pm$ sign in the equation still retains "specific circumstances" tied to individual cases. The next goal is to derive a formulation that eliminates this $\pm$ ambiguity.

To address this, we proceed by taking higher-order derivatives to eliminate the $\pm$ ambiguity.[43] By taking the second derivative with respect to time and space, we have:

$$\frac{\partial^2 \xi}{\partial t^2} = \frac{\partial}{\partial t}\frac{\partial \xi}{\partial t} = \frac{\partial}{\partial t}\left(\pm v \frac{\partial \xi}{\partial \eta}\right) = \pm v \frac{\partial}{\partial \eta}\frac{\partial \xi}{\partial t} = \pm v \frac{\partial}{\partial \eta}\left(\pm v \frac{\partial \xi}{\partial \eta}\right) = v^2 \frac{\partial^2 \xi}{\partial \eta^2}, \tag{2.100}$$

$$\frac{\partial^2 \xi}{\partial x^2} = \frac{\partial}{\partial x}\frac{\partial \xi}{\partial x} = \frac{\partial}{\partial x}\left(\frac{\partial \xi}{\partial \eta}\right) = \frac{\partial}{\partial \eta}\frac{\partial \xi}{\partial x} = \frac{\partial}{\partial \eta}\left(\frac{\partial \xi}{\partial \eta}\right) = \frac{\partial^2 \xi}{\partial \eta^2}.$$

This reveals that the "essential single-variable derivative with respect to η", $\partial^2 \xi / \partial \eta^2$, serves as a common structure, leading to:

$$\left(\frac{\partial^2 \xi}{\partial \eta^2} =\right) \quad \frac{1}{v^2}\frac{\partial^2 \xi}{\partial t^2} = \frac{\partial^2 \xi}{\partial x^2}$$

This is the **wave equation**. To restate its essence: The fact that $\xi(x, t) = \xi(x \pm vt) = \xi(\eta)$ is "essentially a two-variable function dependent on a single variable η" is mathematically expressed as the equality linking time and spatial derivatives at some higher order.

---

[43] Another approach could be squaring both sides to eliminate the $\pm$ sign. However, squaring disrupts linearity, making it unsuitable for describing phenomena where superposition holds, like wave propagation. Therefore, we opt to eliminate the $\pm$ sign by advancing to higher-order derivatives.

Approaches to Explaining the Wave Equation
In most textbooks, the wave equation is introduced by deriving the equation for the vibration of a string, which, in the continuous limit, becomes the wave equation. Some established texts adopt the approach used in this book, where the discussion is based on $\xi(x, t) = \xi(x \pm vt) = \xi(\eta)$. However, these often begin with the statement that "the requirement for waves is equivalent to imposing $\partial^2\xi(x, t)/\partial x \partial t = 0$", which is quite difficult to be understood by beginners [7].

Up to this point, we have explained waves in one dimension, assuming propagation in the x-direction. For three dimensions, we extend this to generalize the wave description by extracting the "direction of wave propagation" from the position vector $\mathbf{r}$ and substituting it into the x-component used in the one-dimensional wave formalism. If the direction of wave propagation is specified using the directional cosine $\hat{\mathbf{u}}$ in three-dimensional space, "extracting the axis of wave propagation from the position vector $\mathbf{r}$" corresponds to taking the projection via the inner product $\mathbf{r} \cdot \hat{\mathbf{u}}$. Thus, we replace the "1D wave component x" with $x \to \mathbf{r} \cdot \hat{\mathbf{u}}$. Consequently,

$$\xi(x, t) = \xi(x \pm v \cdot t)$$

transforms into

$$\xi(\mathbf{r}, t) = \xi(\mathbf{r} \cdot \hat{\mathbf{u}} \pm vt) = \xi(u_x x + u_y y + u_z z \pm vt) = \xi(\eta).$$

From here, we can repeat the same logic used earlier. Following the same reasoning as for the 1D wave equation, we assume that "the partial derivatives with respect to x, y, z, and time will converge at some point as higher-order derivatives with respect to η". We proceed by connecting these derivatives to derive the three-dimensional wave equation.

For the time derivative, the evaluation remains unchanged from the one-dimensional case. However, for the spatial derivatives, such as with respect to x, the change arises due to the "direction cosine coefficient u_x" appearing with each differentiation. This results in:

$$\frac{\partial^2 \xi}{\partial x^2} = u_x^2 \cdot \frac{\partial^2 \xi}{\partial \eta^2}.$$

Similarly, for other directions:

$$\frac{\partial^2 \xi}{\partial y^2} = u_y^2 \cdot \frac{\partial^2 \xi}{\partial \eta^2}, \quad \text{and} \quad \frac{\partial^2 \xi}{\partial z^2} = u_z^2 \cdot \frac{\partial^2 \xi}{\partial \eta^2}.$$

These can be summed up using the fact that the magnitude of the direction cosines satisfies $u_x^2 + u_y^2 + u_z^2 = 1$:

$$\frac{\partial^2 \xi}{\partial x^2} + \frac{\partial^2 \xi}{\partial y^2} + \frac{\partial^2 \xi}{\partial z^2} = (u_x^2 + u_y^2 + u_z^2) \cdot \frac{\partial^2 \xi}{\partial \eta^2} = \frac{\partial^2 \xi}{\partial \eta^2}.$$

Next, equating this result with the time derivative result in Eq. (2.100):

$$\left(\frac{\partial^2 \xi}{\partial \eta^2} =\right) \quad \frac{1}{v^2}\frac{\partial^2 \xi}{\partial t^2} = \frac{\partial^2 \xi}{\partial x^2} + \frac{\partial^2 \xi}{\partial y^2} + \frac{\partial^2 \xi}{\partial z^2},$$

we arrive at the final form:

$$\frac{1}{v^2}\frac{\partial^2 \xi}{\partial t^2} = \nabla^2 \xi. \tag{2.101}$$

This is the **three-dimensional version of the wave equation**. The notation ∇^2 in the above equation is defined as:

$$\nabla^2 = \frac{\partial^2}{\partial x^2} + \frac{\partial^2}{\partial y^2} + \frac{\partial^2}{\partial z^2}.$$

Details about this differential operator will be introduced in the next section. For now, treat it as a simple notation to become familiar with.

To summarize, the dynamics representation of the wave, $\xi(x,t) = f_1(x - v \cdot t) + f_2(x + v \cdot t)$, a quantitative relationship between variables that removes the specificity of individual cases, takes the form of the wave equation:

$$\frac{\partial^2 \xi(x,t)}{\partial t^2} = v^2 \cdot \frac{\partial^2 \xi(x,t)}{\partial x^2}.$$

It is crucial to remain mindful of this methodology of "mathematical representation of dynamics", which abstracts and generalizes the behavior of waves.

2.4.9 Phase Angle and Wave Number

For periodic functions where:

$$f(x + L) = f(x),$$

the "repeating nature" can be naturally visualized as the behavior of a point traveling once around a circle and returning to its starting position. This motion on a circle is governed by the **phase angle** θ. The concept of "shifting x by L" can be projected

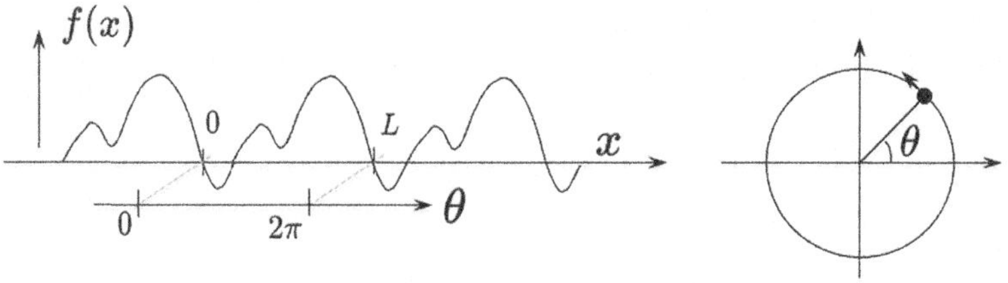

Fig. 2.8 Relationship between phase angle and periodic function. Shifting x by L corresponds to the phase angle $\theta(x)$ completing a full cycle from 0 to 2π

as "the phase angle $\theta(x)$ making a full cycle from 0 to 2π" (Fig. 2.8). Using this projection, the periodic function $f(x)$ can be expressed using the exponential form based on trigonometric functions (Sect. 2.4.4):

$$f(x) \sim \exp\,[i\theta(x)].$$

The proportional relationship for projecting $x = [0, L]$ to $\theta = [0, 2\pi]$ is:

$$\theta(x) = \frac{2\pi}{L} \cdot x =: k \cdot x. \tag{2.102}$$

Using this relationship, the periodicity of $f(x)$ is verified as follows:

$$f(x + L) \sim \exp\,[i\theta(x + L)] = \exp\left[i\frac{2\pi}{L}(x + L)\right]$$

$$= \exp\left[i\frac{2\pi}{L}x\right] \cdot \exp\,[i \cdot 2\pi] = \exp\left[i\frac{2\pi}{L}x\right] = \exp\,[i\theta(x)] \sim f(x).$$

Thus, the periodicity in x is effectively captured as the periodicity of phase angle rotation.

The proportionality constant k in Eq. (2.102) is called the **wave number**. Similarly, for time $t = [0, T]$ projected to $\theta = [0, 2\pi]$, the proportional relationship is:

$$\theta(t) = \frac{2\pi}{T} \cdot t =: \omega \cdot t,$$

where $\omega = 2\pi/T$ is the **angular frequency**.

Understanding Wave Number
When teaching information science students, many easily grasp the concept of angular frequency ω. However, some find the term *wave number k* confusing, likely due to the term's connotation. It is crucial to understand that both ω and k represent just **proportional relationships** projecting intervals such as $t = [0, T]$ or $x = [0, L]$ onto a phase angle cycle $\theta = [0, 2\pi]$. Once this is clear, there is no further reason for confusion. The coefficients are simply chosen such that substituting $x = L$ or $t = T$ cancels the denominator, resulting in $\theta = 2\pi$.

2.4.10 *Phase Angle Representation of Waves*

Wave propagation is expressed through the variable substitution:

$$x \to (x \pm v \cdot t)$$

as discussed in Sect. 2.4.8. Combining this with the periodicity representation:

$$f(x) \sim \exp\left[i\theta(x)\right] \quad , \quad \theta(x) = \frac{2\pi}{L}x,$$

we arrive at the **mathematical representation of periodic waves**. Applying this leads to

$$\theta(x) \to \theta(x \pm v \cdot t) = \theta(x, t),$$

and the corresponding phase angle becomes:

$$\theta(x, t) = \theta(x \pm v \cdot t) = \frac{2\pi}{L}(x \pm v \cdot t) = k \cdot (x \pm v \cdot t),$$

where $k = \frac{2\pi}{L}$ is the wave number.

Here, if we fix the observation point at $x = \bar{x}$ and observe the values of the periodic wave $f(\bar{x}, t)$, this will appear as a periodic variation in time with a period T:

$$f(\bar{x}, t + T) = f(\bar{x}, t). \tag{2.103}$$

At this time, the phase angle is given by:

$$\theta(\bar{x}, t+T) = k \cdot (\bar{x} \pm v \cdot (t+T)) = \theta(\bar{x}, t) \pm k \cdot v \cdot T.$$

For Eq. (2.103) to hold, the remainder $k \cdot v \cdot T$ must equal 2π, ensuring that the phase angles $\theta(\bar{x}, t+T)$ and $\theta(\bar{x}, t)$ are equivalent. This requirement gives $k \cdot v \cdot T = 2\pi$. Comparing $k \cdot v \cdot T = 2\pi$ with $\omega T = 2\pi$, we obtain the relation:

$$\omega = k \cdot v.$$

Using this relationship:

$$\theta(x, t) = \theta(x \pm v \cdot t) = \frac{2\pi}{L}(x \pm v \cdot t) = k\,(x \pm v \cdot t) = k \cdot x \pm \omega \cdot t,$$

we find that a periodic wave propagating in the x-direction takes the form:

$$f(x, t) \sim \exp\left[i \cdot \theta(x, t)\right] = \exp\left[i\,(k \cdot x \pm \omega \cdot t)\right].$$

When expressed precisely, it should be written as "a superposition of all harmonics":

$$f(x, t) = \sum_{\omega}\sum_{k} C_{k,\omega} \cdot \exp\left[i\,(k \cdot x \pm \omega \cdot t)\right],$$

which represents the wave-extended version of the Fourier series expansion in Eq. (2.88).

In the three-dimensional case, as described at the end of Sect. 2.4.8, we consider the projection of the wave propagation direction cosine $\hat{\mathbf{u}}$ onto the x-axis. By substituting $x = \hat{\mathbf{u}} \cdot \mathbf{r}$, the wave can be expressed as:

$$f(\mathbf{r}, t) \sim \exp\left[i\left(k \cdot \hat{\mathbf{u}} \cdot \mathbf{r} \pm \omega \cdot t\right)\right].$$

Here, defining the **wave vector k** as:

$$\mathbf{k} := k \cdot \hat{\mathbf{u}} \qquad [*] \tag{2.104}$$

allows the wave to be represented in three dimensions as:

$$f(\mathbf{r}, t) \sim \exp\left[i\,(\mathbf{k} \cdot \mathbf{r} \pm \omega \cdot t)\right].$$

Note that the *direction of the wave vector* **k** represents the **direction of wave propagation**. Among the signs, the negative sign corresponds to a wave propagating in the positive direction. Thus, a wave propagating in the positive direction can be expressed as:

$$f(\mathbf{r}, t) = \sum_{\omega} \sum_{\mathbf{k}} C_{\mathbf{k},\omega} \cdot \exp\left[i\left(\mathbf{k} \cdot \mathbf{r} - \omega \cdot t\right)\right] \tag{2.105}$$

Why This Section Was Written
When attempting to discuss waves with students, especially those majoring information sciences or other backgrounds without prior exposure to general physics, I have repeatedly encountered confusion at the appearance of the factor $e^{i(\mathbf{k}\cdot\mathbf{r}-\omega\cdot t)}$. This repeated experience served as the motivation for writing this section.

2.5 Field Analysis

The spatial distributions of quantities such as temperature, altitude, or pressure can be abstracted as functions that assign scalar values to each point in space $\mathbf{r} = (x, y, z)$. These are described as **scalar fields** and are represented as $\phi(\mathbf{r}) = \phi(x, y, z)$. Similarly, spatial distributions of quantities such as wind direction or heat flow can be abstracted as vector functions $\mathbf{A}(\mathbf{r}) = \mathbf{A}(x, y, z)$, which are defined at each point in space $\mathbf{r} = (x, y, z)$. These are referred to as **vector fields**.

Gradual Introduction to Vector Fields
This section begins with an elementary and intuitive introduction to these concepts to familiarize readers with the framework. Later, the notion of "fields defined over coordinate space" will be revisited and updated. Reached at this stage, fields are classified based on their transformations under coordinate transformations (Sect. 6.2.1). A **scalar field** is defined as a field invariant under coordinate transformations, while a **vector field** is one that transforms according to the same rules as the components of position vectors. Following these definitions, other fields such as tensor fields and spinor fields are introduced (Sect. 7).

Since vector fields and scalar fields are associated with each point in space, operational concepts such as "**Integrating** these field quantities **along a given trajectory**, or **over a given region**" naturally arise. This section discusses how to formalize these operations.

In response to the concept of **field integration**, the concept of **field differentiation** also arises. Field derivatives such as $\nabla\phi(\mathbf{r})$, $\nabla\mathbf{A}(\mathbf{r})$, or $\nabla \times \mathbf{A}(\mathbf{r})$ appear frequently not only in physics and engineering but also in various scientific fields, including numerical optimization, which forms the foundation of machine learning. This section provides a natural introduction to the meaning and significance of these derivatives.

2.5.1 Trajectory Description and Line Integrals

To introduce the operational concept of "integrating field quantities along a trajectory C in space", we first represent the trajectory C in three-dimensional space as:

$$\mathbf{r}(t) = \begin{pmatrix} x(t) \\ y(t) \\ z(t) \end{pmatrix},$$

which is a "position vector varying with a **single parameter**". As the parameter t changes continuously, the position indicated by the position vector also changes continuously, tracing out the trajectory.

Expanding $\mathbf{r}(t + dt)$ using the Taylor series (Sect. 2.1.1):

$$\mathbf{r}(t + dt) \approx \mathbf{r}(t) + \frac{d\mathbf{r}}{dt}dt\ , \ \therefore \quad d\mathbf{r} = [\mathbf{r}(t + dt) - \mathbf{r}(t)] = \frac{d\mathbf{r}}{dt}dt, \tag{2.106}$$

we see that $d\mathbf{r}$ geometrically represents the vector connecting the "tip of $\mathbf{r}(t)$" to the "tip of $\mathbf{r}(t + dt)$". It corresponds to the "tangent along the trajectory", as shown in Fig. 2.9a. Therefore, the "direction as a vector" of $d\mathbf{r}$ or $d\mathbf{r}/dt$ can be confirmed to point in the tangential direction of the trajectory.

Thus,

$$\mathbf{A}(\mathbf{r}) \cdot d\mathbf{r} = \mathbf{A}(\mathbf{r}) \cdot \frac{d\mathbf{r}}{dt}\, dt$$

is interpreted as the "component of the vector field $\mathbf{A}(\mathbf{r})$ projected along the infinitesimal line element $d\mathbf{r}$". The integral of this projected component, which is **summed along the trajectory** (as denoted on the left-hand side), is defined as:

$$\begin{aligned} \int_C \mathbf{A}(\mathbf{r}) \cdot d\mathbf{r} &:= \int_{t_1}^{t_2} \begin{pmatrix} A_x(t) \\ A_y(t) \\ A_z(t) \end{pmatrix} \cdot \begin{pmatrix} dx(t) \\ dy(t) \\ dz(t) \end{pmatrix} \\ &= \int_{t_1}^{t_2} \left\{ A_x(t)dx(t) + A_y(t)dy(t) + A_z(t)dz(t) \right\} \end{aligned}$$

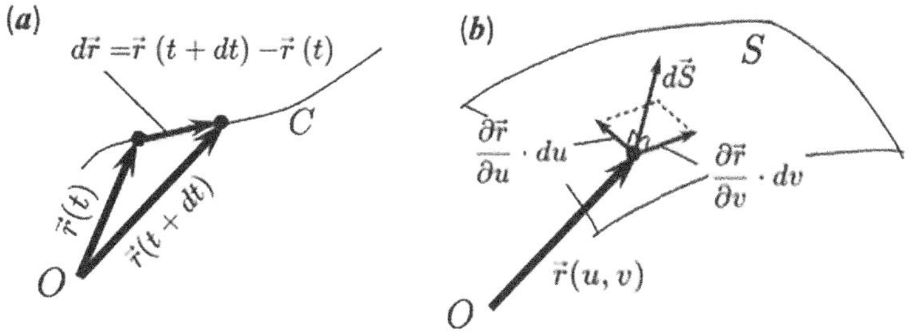

Fig. 2.9 Trajectories and surfaces are parametrized as shown in this figure. The vector $d\mathbf{r}$, connecting the "tip of $\mathbf{r}(t)$" and the "tip of $\mathbf{r}(t+dt)$", represents the *tangent along the trajectory* (**a**). The vectors $\left(\frac{\partial \mathbf{r}}{\partial u}\right) du$ and $\left(\frac{\partial \mathbf{r}}{\partial v}\right) dv$ are *tangent vectors on the surface* along their respective directions. Their cross product $d\mathbf{S}$ represents the *normal vector to the surface*, perpendicular to these two tangent vectors

$$= \int_{t_1}^{t_2} \left\{ A_x(t)\frac{dx}{dt}dt + A_y(t)\frac{dy}{dt}dt + A_z(t)\frac{dz}{dt}dt \right\}$$
$$= \int_{t_1}^{t_2} \left\{ A_x(t)\frac{dx}{dt} + A_y(t)\frac{dy}{dt} + A_z(t)\frac{dz}{dt} \right\} dt.$$

This is the integral value for real functions (rightmost expression) defined by this computation. Here, t_1 and t_2 are the parameter values corresponding to the start and end points of the trajectory C.

Next, let us consider the **line integral of the scalar field** $\phi(\mathbf{r})$ along a trajectory. This can be expressed using the infinitesimal line element $ds = |d\mathbf{r}|$ along the trajectory as:

$$\phi(\mathbf{r}) \cdot |d\mathbf{r}| = \phi(\mathbf{r}) \cdot ds = \phi(\mathbf{r}) \cdot dt \left| \frac{d\mathbf{r}}{dt} \right|,$$

which can be computed as the following integral:

$$\int_C \phi(\mathbf{r}) \cdot ds := \int_{t_1}^{t_2} \phi(x(t), y(t), z(t)) \cdot dt \left| \frac{d\mathbf{r}}{dt} \right|,$$

defined as the integral computation for real functions.

2.5.2 *Description of Surfaces, Surface Integrals*

Next, we consider the formalization of the algorithm for integrating scalar fields and vector fields over a surface S (surface integrals). By extending the concept of a *one-*

parameter description of a one-dimensional trajectory, a surface can be described as the region swept by the tip of a position vector $\mathbf{r}(u, v)$ defined by **two parameters**.

If we fix one parameter and vary the other, similar to the discussion for one-dimensional trajectories [Eq. (2.106)], the variations $\left(\frac{\partial \mathbf{r}}{\partial u}\right) du$ and $\left(\frac{\partial \mathbf{r}}{\partial v}\right) dv$ represent **tangent vectors on the surface** along each respective direction (two vectors lying on the tangent plane). This is illustrated in Fig. 2.9b. Therefore, their cross product

$$\frac{\partial \mathbf{r}}{\partial u} du \times \frac{\partial \mathbf{r}}{\partial v} dv = \left(\frac{\partial \mathbf{r}}{\partial u} \times \frac{\partial \mathbf{r}}{\partial v} \right) dudv$$

yields a *normal vector to the surface* perpendicular to these two tangent vectors (Sect. 2.2.5). We then define:

$$d\mathbf{S} = \left(\frac{\partial \mathbf{r}}{\partial u} \times \frac{\partial \mathbf{r}}{\partial v} \right) dudv, \quad , \quad dS = |d\mathbf{S}| = \left| \frac{\partial \mathbf{r}}{\partial u} \times \frac{\partial \mathbf{r}}{\partial v} \right| dudv.$$

Here, $d\mathbf{S}$ is called the **surface element vector**. The magnitude $dS = |d\mathbf{S}|$ represents the *infinitesimal surface area element*, and the direction of $d\mathbf{S}$ points along the *normal direction to the surface*.

We have now formulated the infinitesimal surface element $d\mathbf{S}$ for integration over a surface. When a point $\mathbf{r}(u, v)$ on the surface S varies according to two parameters (u, v), the scalar field $\phi(\mathbf{r})$ or the vector field $\mathbf{A}(\mathbf{r})$ changes its value accordingly. Therefore, the term $\mathbf{A}(\mathbf{r}) \cdot d\mathbf{S}$ can be interpreted as the *component of the vector field* $\mathbf{A}(\mathbf{r})$ *projected onto the normal direction of the infinitesimal surface element*. The integral of this projected component over the surface, defined as:

$$\int_S \mathbf{A}(\mathbf{r}) \cdot d\mathbf{S} := \int_{u_1}^{u_2} \int_{v_1}^{v_2} \mathbf{A}(\mathbf{r}(u, v)) \cdot \left(\frac{\partial \mathbf{r}}{\partial u} \times \frac{\partial \mathbf{r}}{\partial v} \right) dudv,$$

is referred to as the **surface integral of the vector field**.

On the other hand, for a scalar field, the product of the scalar field value and the magnitude of the infinitesimal surface element:

$$\phi(\mathbf{r}) \cdot |d\mathbf{S}| = \phi(\mathbf{r}) \cdot dS = \phi(\mathbf{r}) \cdot \left| \frac{\partial \mathbf{r}}{\partial u} \times \frac{\partial \mathbf{r}}{\partial v} \right| dudv$$

is summed up to define the integral:

$$\int_S \phi(\mathbf{r}) dS := \int_{u_1}^{u_2} \int_{v_1}^{v_2} \phi(\mathbf{r}) \left| \frac{\partial \mathbf{r}}{\partial u} \times \frac{\partial \mathbf{r}}{\partial v} \right| dudv.$$

This is called the **surface integral of the scalar field**.

Basis Vectors Can Be Defined via Partial Derivatives
Looking at Fig. 2.9b, the terms $(\partial\mathbf{r}/\partial u)du$ and $(\partial\mathbf{r}/\partial v)dv$ for $\mathbf{r}(u, v)$ appear to act as new basis vectors on the surface. Recall that in the case of spherical coordinates $\mathbf{r}(r, \theta, \phi)$, the directions of variation given by $(\partial\mathbf{r}/\partial r)$, $(\partial\mathbf{r}/\partial\theta)$, and $(\partial\mathbf{r}/\partial\phi)$ correspond to $\{\mathbf{e}_r, \mathbf{e}_\theta, \mathbf{e}_\phi\}$. Using Einstein's summation convention, for example, $\mathbf{e}_\theta = (\partial\mathbf{r}/\partial\theta) = (\partial x_l/\partial\theta)\cdot\mathbf{e}_l$. More generally, if we express the parametrization as $\mathbf{r}(\{x'_j\})$, the transformation rule for the basis vectors can be written as $\mathbf{e}'_j = (\partial x_l/\partial x'_j)\cdot\mathbf{e}_l$. Comparing this with Eq. (2.48), $\mathbf{e}'_j = \hat{A}\cdot\mathbf{e}_j = a_{lj}\cdot\mathbf{e}_l$, we find that $a_{lj} = (\partial x_l/\partial x'_j)$. As mentioned in the footnote of Sect. 2.3.3, quantities that transform in the same way as basis vectors are called **covariant** quantities. Therefore, $A'_j = (\partial x_l/\partial x'_j)\cdot A_l$ defines a covariant quantity [4]. Understanding this brings you to the threshold of the field of **differential geometry**. In this context, we consider the **tangent space** defined by operations of partial differentiation at each point in space, paired with the vector space. If such concepts are introduced with ready-made definitions, beginners may feel confused. However, the explanation above provides a natural understanding of why we consider operations involving partial derivatives with respect to parameters.

2.5.3 Differentiation of Scalar Fields

Having discussed the integration of fields, we now move on to **differentiation of fields**. We will first discuss the differentiation of scalar fields, followed by the differentiation of vector fields.

Consider a scalar function $\phi(\mathbf{r}) = \phi(x, y, z)$ defined in three-dimensional space with a position vector $\mathbf{r} = (x, y, z)$. Compared to single-variable calculus, there are more variations of partial derivatives:

$$\frac{\partial\phi(\mathbf{r})}{\partial x}, \quad \frac{\partial\phi(\mathbf{r})}{\partial y}, \quad \frac{\partial\phi(\mathbf{r})}{\partial z}$$

These can be expressed as:

$$\begin{pmatrix}\partial\phi(\mathbf{r})/\partial x\\ \partial\phi(\mathbf{r})/\partial y\\ \partial\phi(\mathbf{r})/\partial z\end{pmatrix} = \begin{pmatrix}\partial/\partial x\\ \partial/\partial y\\ \partial/\partial z\end{pmatrix}\cdot\phi(\mathbf{r}) = \nabla\phi(\mathbf{r}) \tag{2.107}$$

The operator ∇ is called the **nabla.**[44]

This **differentiation of scalar fields**, $\nabla\phi(\mathbf{r})$, describes the **direction of the normal to the equipotential surface** and also the **direction of the maximum gradient** of $\phi(\mathbf{r})$. The following two subsections will explain these ideas further.

2.5.4 Normal Vector of an Equipotential Surface

As described in Sect. 2.5.2, the formulation of a surface is represented by the position vector that "sweeps the surface at its tip":

$$\mathbf{r} = (x, y, z(x, y)) = \mathbf{r}(x, y) \tag{2.108}$$

with a two-parameter dependency on (x, y). Recalling what was described in Sect. 2.5.2, the normal vector $\mathbf{n}$ is given by:

$$\mathbf{n} \propto \left(\frac{\partial \mathbf{r}}{\partial x} \times \frac{\partial \mathbf{r}}{\partial y}\right).$$

From Eq. (2.108),

$$\frac{\partial \mathbf{r}}{\partial x} = \left(1, 0, \frac{\partial z(x, y)}{\partial x}\right) \quad , \quad \frac{\partial \mathbf{r}}{\partial y} = \left(0, 1, \frac{\partial z(x, y)}{\partial y}\right).$$

Taking their cross product [(see Eq. (2.14)], we get

$$\left(\frac{\partial \mathbf{r}}{\partial x} \times \frac{\partial \mathbf{r}}{\partial y}\right) = \begin{vmatrix} \mathbf{e}_x & \mathbf{e}_y & \mathbf{e}_z \\ 1 & 0 & \partial z(x, y)/\partial x \\ 0 & 1 & \partial z(x, y)/\partial y \end{vmatrix} = (-\partial z(x, y)/\partial x, -\partial z(x, y)/\partial y, 1).$$

Thus, the normal vector $\mathbf{n}$ is:

$$\mathbf{n} \propto \left(-\frac{\partial z(x, y)}{\partial x}, -\frac{\partial z(x, y)}{\partial y}, 1\right). \tag{2.109}$$

[44] As noted in the introductory column of this section, in more advanced definitions of vectors, scalars, vectors, and tensors are classified and defined based on how they transform under changes of coordinates (Sect. 6.2.1). According to this definition, ∇ qualifies as a vector, justifying its treatment as a typical vector [1]. It is important to note that simply having a triplet of components does not automatically justify expressing it as a vector.

To evaluate $\partial z\,(x, y)\,/\partial x$ and $\partial z\,(x, y)\,/\partial y$ that appear here, we differentiate both sides of the equipotential surface equation:

$$\phi\,(x, y, z\,(x, y)) = C$$

with respect to x and y, respectively:

$$\frac{\partial \phi}{\partial x} + \frac{\partial \phi}{\partial z}\frac{\partial z}{\partial x} = 0 \quad \rightarrow \quad \frac{\partial z}{\partial x} = -\frac{\partial \phi}{\partial x}\left(\frac{\partial \phi}{\partial z}\right)^{-1},$$

$$\frac{\partial \phi}{\partial y} + \frac{\partial \phi}{\partial z}\frac{\partial z}{\partial y} = 0 \quad \rightarrow \quad \frac{\partial z}{\partial y} = -\frac{\partial \phi}{\partial y}\left(\frac{\partial \phi}{\partial z}\right)^{-1}.$$

Thus, we obtain expressions for $\partial z\,(x, y)\,/\partial x$ and related derivatives (Sect. 2.4.7). These expressions for $\partial z/\partial x$ and $\partial z/\partial y$ can be substituted into Eq. (2.109), yielding the normal vector to the equipotential surface $\phi(x, y, z) = C$:

$$\mathbf{n} \propto \left(-\frac{\partial z(x, y)}{\partial x}, -\frac{\partial z(x, y)}{\partial y}, 1\right) = \left(\frac{\partial \phi}{\partial x}\left(\frac{\partial \phi}{\partial z}\right)^{-1}, \frac{\partial \phi}{\partial y}\left(\frac{\partial \phi}{\partial z}\right)^{-1}, 1\right)$$

$$\propto \left(\frac{\partial \phi}{\partial x}, \frac{\partial \phi}{\partial y}, \frac{\partial \phi}{\partial z}\right) = \boldsymbol{\nabla}\phi.$$

Thus, we obtain the result that: $\boldsymbol{\nabla}\phi(\mathbf{r})$ **gives the normal vector n of the equipotential surface** $\phi\,(\mathbf{r}) = C$.

2.5.5 *Meaning as Gradient Direction*

In the previous subsection, we explained that $\boldsymbol{\nabla}\phi(\mathbf{r})$ gives the normal to an equipotential surface. Now, consider a curve that passes through successive equipotential surfaces and imagine the change in potential along that path. In two dimensions, these equipotential surfaces correspond to contour lines on a map. Recall that "contour lines spaced closely together indicate steep gradients". Similarly, in three-dimensional space, where equipotential surfaces are dense, changes are rapid, and where they are sparse, changes are minimal. If a path traverses a region with dense equipotential surfaces, the potential change is significant, indicating a steep slope. This section derives the conclusion that: "If you move in the direction of $\boldsymbol{\nabla}\phi\,(\mathbf{r})$ at a given point $\mathbf{r}$, you follow the direction where the **potential changes most rapidly**".

Let the path be defined as $\mathbf{r}(t) = (x(t), y(t), z(t))$.[45] The change in the scalar field $\phi(x, y, z)$ along this path is:[46]

$$\frac{d\phi(x(t), y(t), z(t))}{dt} = \frac{\partial\phi(x, y, z)}{\partial x}\frac{dx(t)}{dt} + \frac{\partial\phi(x, y, z)}{\partial y}\frac{dy(t)}{dt} + \frac{\partial\phi(x, y, z)}{\partial z}\frac{dz(t)}{dt}.$$

This can be expressed as the inner product with $\nabla\phi$:

$$\frac{d\phi\,(\mathbf{r}\,(t))}{dt} = (\nabla\phi)\cdot\frac{d\mathbf{r}\,(t)}{dt} \tag{2.110}$$

We will consider this equation in the following discussion.

Now, $\frac{d\mathbf{r}(t)}{dt}$ is the tangent vector to the path (Fig. 2.9a). To maximise the potential change $d\phi/dt$ in Eq. (2.110), the direction of $\frac{d\mathbf{r}(t)}{dt}$ should be aligned with $\nabla\phi$.

Next, let's evaluate the **magnitude of the gradient along this direction**. The gradient is defined as the rate of change when moving by a unit vector.[47] Since the maximum gradient direction is aligned with $\nabla\phi$, the unit vector in this direction is:

$$\left.\frac{d\mathbf{r}(t)}{dt}\right|_{\text{max/unit}} = \frac{\nabla\phi}{|\nabla\phi|}.$$

Substituting into Eq. (2.110), the gradient magnitude is:

$$\left(\frac{d\phi(\mathbf{r}(t))}{dt}\right)_{\text{max/unit}} = \nabla\phi\cdot\frac{\nabla\phi}{|\nabla\phi|} = |\nabla\phi|.$$

Therefore, the conclusion is: **"The direction of $\nabla\phi(\mathbf{r})$ indicates the direction of the steepest potential change at point r, and its magnitude represents the gradient (rate of potential change) along a unit vector in that direction".** This interpretation gives rise to the term **"gradient of the scalar field $\phi(\mathbf{r})$"** for $\nabla\phi(\mathbf{r})$.

[45] Paths or trajectories are described as position vectors defined by a single parameter t (Sect. 2.5.1).

[46] Recall the chain rule for partial derivatives discussed in Sect. 2.4.7.

[47] In middle school, when learning about the linear function $y = ax + b$, the slope a was defined as "the increase when x advances by 1". Drawing the graph, a is indeed "the increase when x advances by 1".

2.5.6 Differentiation of Vector Fields

In the previous sections, we discussed the differentiation of scalar fields and their meanings. Next, we consider the differentiation of a vector field:

$$\mathbf{A}(\mathbf{r}) = \mathbf{A}(x, y, z) = \begin{pmatrix} A_x(x, y, z) \\ A_y(x, y, z) \\ A_z(x, y, z) \end{pmatrix}.$$

In this case, the derivatives have more variations:

$$\begin{pmatrix} \partial_x A_x & \partial_y A_x & \partial_z A_x \\ \partial_x A_y & \partial_y A_y & \partial_z A_y \\ \partial_x A_z & \partial_y A_z & \partial_z A_z \end{pmatrix} = \{\partial_\alpha A_\beta\}, \tag{2.111}$$

resulting in nine different terms.[48] These nine terms can be rearranged into two quantities:[49]

$$\nabla \cdot \mathbf{A}(\mathbf{r}) = \left[\partial_x A_x + \partial_y A_y + \partial_z A_z\right],$$

$$\nabla \times \mathbf{A}(\mathbf{r}) = \begin{pmatrix} \partial_y A_z - \partial_z A_y \\ \partial_x A_z - \partial_z A_x \\ \partial_x A_y - \partial_y A_x \end{pmatrix}.$$

To construct these two quantities, all nine components are required, meaning no information is reduced compared to Eq. (2.111). However, instead of interpreting the nine elements individually, they acquire meaning collectively as the pair $\nabla \cdot \mathbf{A}(\mathbf{r})$ and $\nabla \times \mathbf{A}(\mathbf{r})$.

2.5.7 Interpreting the Fundamental Theorem of Calculus

To explain the meaning of $\nabla \cdot \mathbf{A}(\mathbf{r})$ and $\nabla \times \mathbf{A}(\mathbf{r})$, we start by revisiting the fundamental theorem of calculus, which appeared in high school:

$$\int_a^b \frac{df(x)}{dx} dx = [f(b) - f(a)] = [f(x)]_a^b. \tag{2.112}$$

[48] Here, $\partial/\partial x$ is abbreviated as ∂_x, and similarly for other partial derivatives.

[49] The symbols $\times$ and $\cdot$ denote the vector cross product and dot product, respectively.

In this equation, the right-hand side $[f(x)]_a^b$ represents the values at the boundaries of the integration interval $[a, b]$; it captures information at the **edges of the region**.

When a region is denoted by Ω, the "boundary of the region Ω" is written as $\partial\Omega$.

- $\partial(V) = S$; the boundary of a 3D volume is its surface (2D).
- $\partial(S) = C$; the boundary of a 2D surface is its edge curve (1D).
- $\partial(L) = P$; the boundary of a 1D line segment is its endpoints (0D).

These examples help visualize the concept clearly. Using this notation, the right-hand side of the fundamental theorem can be expressed as:

$$[f(x)]_a^b = \int_a^b f(x)d(\partial x) = [+f(b) + (-f(a))] .$$

The integral $\int_a^b [\cdots]d(\partial x)$ represents a sum over the "edges ∂x" of the line segment, i.e., the points a and b, with weights assigned: positive at point b and negative at point a. This can be understood by recalling the "direction of surface vectors" explained in Fig. 2.9b. If we interpret points a and b as "surfaces perpendicular to the x-axis", the outward-facing normal at b aligns with the positive x-axis (positive sign), while at a it points in the opposite direction (negative sign).

Thus, the fundamental theorem of calculus can be rewritten as

$$\int_a^b \frac{df(x)}{dx}dx = \int_a^b f(x)d(\partial x). \tag{2.113}$$

This equation expresses the idea that if we integrate the local change $df(x)/dx$ over a region, the result is the total change for the entire region. This total change corresponds to the net difference in flow across the boundaries, reflecting a **quantitative conservation law**.

2.5.8 Gauss's and Stokes's Theorem

For the one-dimensional case, we rewrote the fundamental theorem of calculus as in Eq. (2.113). Extending this, we can express it as

$$\int_{\partial\Omega} f(\Omega)\, d(\partial\Omega) = \int_\Omega \frac{df(\Omega)}{d\Omega} d\Omega. \tag{2.114}$$

This type of quantitative conservation law can generally apply to higher-dimensional regions, such as two-dimensional or three-dimensional domains. This equation can be viewed as a tool for **transforming the integration domain** $\partial\Omega \to \Omega$.

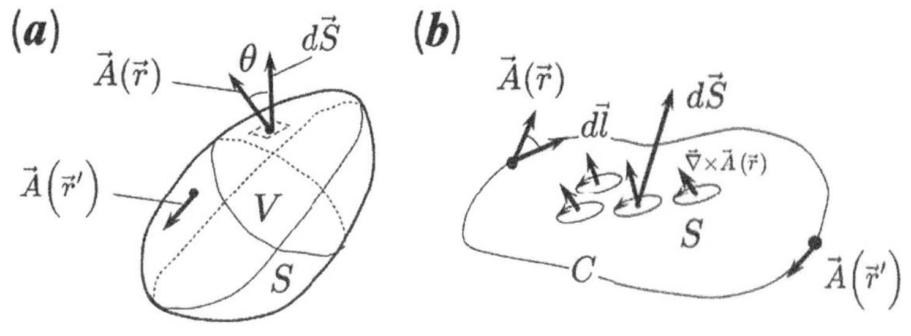

Fig. 2.10 When recalling Gauss's theorem or Stokes's theorem, the first images to bring to mind are as follows: For Gauss's theorem (**a**), imagine a volume region V enclosed by a closed surface S. For Stokes's theorem (**b**), visualize a surface region S enclosed by a closed curve C

In Sect. 2.5.1 and Sect. 2.5.2, we introduced the line and surface integrals of a vector field $\mathbf{A}(\mathbf{r})$:

$$\oint_c d\mathbf{l} \cdot \mathbf{A}(\mathbf{r}) \quad , \quad \int_S d\mathbf{S} \cdot \mathbf{A}(\mathbf{r}).$$

Applying Eq. (2.114) with $c = \partial S$ and $S = \partial V$, these integrals become:

$$\oint_{c=\partial S} d\mathbf{l} \cdot \mathbf{A}(\mathbf{r}) = \int_S d\mathbf{S} \cdot \mathbf{F}(\{\partial_\alpha A_\beta\}) \tag{2.115}$$

$$\int_{S=\partial V} d\mathbf{S} \cdot \mathbf{A}(\mathbf{r}) = \int_V dV \cdot G(\{\partial_\alpha A_\beta\}). \tag{2.116}$$

These expressions indicate that the integrals on the left-hand side can be evaluated as integrals over a region Ω of some functions $\mathbf{F}$ or G, which are constructed from the differential terms $\{\partial_\alpha A_\beta\}$.[50] Note that the left-hand sides of both equations are scalar quantities. In Eq. (2.115), since $d\mathbf{S}$ is a vector, $\mathbf{F}$ must be a vector function, ensuring their dot product results in a scalar. In Eq. (2.116), dV is a scalar, so G must be a scalar function.

The line integral on the left-hand side of Eq. (2.115) captures the "circulation of the vector field $\mathbf{A}(\mathbf{r})$ along the closed curve $c = \partial S$", as illustrated in Fig. 2.10b. The surface integral on the left-hand side of Eq. (2.116) captures the "flux of the vector field $\mathbf{A}(\mathbf{r})$ through the closed surface $S = \partial V$", as shown in Fig. 2.10a. Thus, the left-hand sides of Eqs. (2.115) and (2.116) represent "global circulation or global flux captured at the boundary $\partial\Omega$". These expressions are equal to the right-hand sides, which represent the "sum of local quantities inside the region Ω".

[50] The symbol $\oint_c$ denotes a line integral over a closed path.

Therefore, the integrands $\mathbf{F}(\{\partial_\alpha A_\beta\})$ and $G(\{\partial_\alpha A_\beta\})$ represent the **local circulation** and **local flux**, respectively. It can be shown that these functions $\mathbf{F}$ and G take the form:

$$\mathbf{F}\left(\{\partial_\alpha A_\beta\}\right) = \begin{pmatrix} \partial_y A_z - \partial_z A_y \\ \partial_x A_z - \partial_z A_x \\ \partial_x A_y - \partial_y A_x \end{pmatrix} = \nabla \times \mathbf{A}(\mathbf{r}),$$

$$G\left(\{\partial_\alpha A_\beta\}\right) = \left[\partial_x A_x + \partial_y A_y + \partial_z A_z\right] = \nabla \cdot \mathbf{A}(\mathbf{r}).$$

Rewriting these expressions gives:

$$\oint_C d\mathbf{l} \cdot \mathbf{A}(\mathbf{r}) = \int_S d\mathbf{S} \cdot [\nabla \times \mathbf{A}(\mathbf{r})],$$

$$\int_S d\mathbf{S} \cdot \mathbf{A}(\mathbf{r}) = \int_V dV\, [\nabla \cdot \mathbf{A}(\mathbf{r})]. \tag{2.117}$$

These are known as **Stokes's Theorem** and **Gauss's Theorem**, respectively. The expression $\mathbf{F} = \nabla \times \mathbf{A}\,(\mathbf{r})$ represents the "microscopic vorticity of the vector field $\mathbf{A}\,(\mathbf{r})$ at position $\mathbf{r}$" and is referred to as **rotation**. Similarly, $G = \nabla \cdot \mathbf{A}\,(\mathbf{r})$ represents the "source density of the vector field $\mathbf{A}\,(\mathbf{r})$ at position $\mathbf{r}$" and is referred to as **divergence**.

Proofs of these theorems can be found in many standard textbooks [1, 8], so we will avoid reproducing the detailed proofs here.[51] The outline of the proof involves explicitly writing the meanings of the line integrals and surface integrals in component form, and applying the fundamental theorem of calculus (2.112) to derive the results. For example, in the proof of Gauss's theorem (Eq. (2.117)), we have:

$$\int_V \nabla \cdot \mathbf{A}(\mathbf{r})\, dV = \int_V \left[\frac{\partial A_x(x, y, z)}{\partial x} + \frac{\partial A_y(x, y, z)}{\partial y} + \frac{\partial A_z(x, y, z)}{\partial z}\right] dxdydz.$$

For the first term, for instance:

$$\int_V \left[\frac{\partial A_x(x, y, z)}{\partial x}\right] dxdydz = \int dydz \underline{\int \frac{\partial A_x(x, y, z)}{\partial x} dx}.$$

By applying the fundamental theorem of calculus to the underlined part, the relationship in Eq.(contour) is derived.

Henceforth, Stokes's theorem and Gauss's theorem will frequently appear when making **inferences using mathematical formulas**. Therefore, it is essential to

[51] Most traditional textbooks illustrate the changes described by $\nabla \cdot \mathbf{A}(\mathbf{r})$ and $\nabla \times \mathbf{A}(\mathbf{r})$ with figures, explaining how they correspond to divergence and rotation, respectively [8]. On the other hand, books that first introduce the integral theorems and then derive $\nabla \cdot \mathbf{A}(\mathbf{r})$ and $\nabla \times \mathbf{A}(\mathbf{r})$, as this book does, are relatively a few [1].

memorize and write them down while keeping their meanings in mind. For both Gauss's theorem and Stokes's theorem, recall the visualizations of line integrals and surface integrals introduced in Sects. 2.5.1 and 2.5.2. Begin by conjuring images like those in Fig. 2.10.

First, recall that in all cases, the integral under consideration is an "integration of the vector field $\mathbf{A}(\mathbf{r})$":

$$I = \int \mathbf{A}(\mathbf{r}) \cdot [\ldots] .$$

Next, recall that vector field integrals can be classified as either line integrals or surface integrals. Since the integral value I is a scalar, it is essential to consider that "a scalar quantity is formed by taking the inner product with the vector field $\mathbf{A}(\mathbf{r})$". With this in mind, we write down the two possible forms:

$$I = \begin{cases} \displaystyle\int d\mathbf{S} \cdot \mathbf{A}(\mathbf{r}) \\[2ex] \displaystyle\int d\mathbf{l} \cdot \mathbf{A}(\mathbf{r}) \end{cases}$$

The first (upper) expression corresponds to Gauss's theorem, while the second (lower) corresponds to Stokes' theorem. Following this, sketch a closed surface and a closed curve, as illustrated in Fig. 2.10, to proceed with the considerations.

How to Write Down Gauss's Theorem

Start by sketching on a piece of paper the situation of "a vector field flowing out from within a closed surface", as illustrated in Fig. 2.10a. Next, recall the idea of "capturing the outflow on the surface", and write down the left-hand side of the equation:

$$\int_S d\mathbf{S} \cdot \mathbf{A}(\mathbf{r}) = \int_V dV \left[\nabla \cdot \mathbf{A}(\mathbf{r})\right].$$

The left-hand side represents "the sum of the components of the vector field $\mathbf{A}(\mathbf{r})$ projected in the direction of the surface element vector $d\mathbf{S}$, summed over the entire closed surface". Pay attention to the directions of $\mathbf{A}(\mathbf{r})$ and $\mathbf{A}\left(\mathbf{r}'\right)$ at different positions, as shown in Fig. 2.10a. In the case of $\mathbf{A}(\mathbf{r})$, the projection in the $d\mathbf{S}$ direction is well counted, whereas in the case of $\mathbf{A}\left(\mathbf{r}'\right)$, the vector field flows along the surface and is orthogonal to $d\mathbf{S}$, so it contributes nothing. Additionally, if $\mathbf{A}(\mathbf{r})$ is directed inward toward the closed surface S, the dot product gives a negative value, which is counted negatively. From this, it becomes clear that "counting the projection in the $d\mathbf{S}$ direction" corresponds to "counting the total outflow through the closed surface S".

Next, write down the right-hand side, being equated to the left-hand side. It is important to **remember in advance** that $\nabla \cdot \mathbf{A}(\mathbf{r})$ is referred to as divergence, meaning local outflow. By understanding and reciting the principle that "the total outflow through a surface is the sum of the local outflows inside", writing down the volume integral of $\nabla \cdot \mathbf{A}(\mathbf{r})$ yields Gauss's integral theorem.

How to Write Down Stokes's Theorem

Start by sketching on a piece of paper the situation where "a swirling vector field $\mathbf{A}(\mathbf{r})$ is enclosed by a closed curve C", as shown in Fig. 2.10b. Next, recall the idea of "capturing the vorticity along the closed curve", and write down the left-hand side of the equation:

$$\oint_C d\mathbf{l} \cdot \mathbf{A}(\mathbf{r}) = \int_S d\mathbf{S} \cdot [\nabla \times \mathbf{A}(\mathbf{r})].$$

The left-hand side represents "the sum of contributions from the vector field $\mathbf{A}(\mathbf{r})$ projected onto the tangential direction $d\mathbf{l}$ of the closed curve, integrated over one closed loop". Pay attention to the directions of $\mathbf{A}(\mathbf{r})$ and $\mathbf{A}(\mathbf{r}')$ at different positions, as shown in Fig. 2.10b. In the case of $\mathbf{A}(\mathbf{r}')$, the projection onto the $d\mathbf{l}$ direction is mostly counted, whereas for $\mathbf{A}(\mathbf{r})$, the vector field tends to be more orthogonal to the closed curve at that location, resulting in a smaller contribution. Additionally, if $\mathbf{A}(\mathbf{r})$ is oriented opposite to the $d\mathbf{l}$ direction, the dot product gives a negative value, which is counted negatively. From this, it becomes clear that "counting the projection in the $d\mathbf{l}$ direction" corresponds to "counting the total contribution of vorticity along the closed curve C".

Next, write down the right-hand side, being equated to the left-hand side. It is important to **remember in advance** that $\nabla \times \mathbf{A}(\mathbf{r})$ is referred to as rotation, meaning local vorticity. By reciting the principle that "the total vorticity remaining at the edge of the closed curve is the sum of the local vorticity $\nabla \times \mathbf{A}(\mathbf{r})$ over the enclosed surface", and writing down the surface integral, we obtain Stokes' integral theorem. Note also that the local vorticity $\nabla \times \mathbf{A}(\mathbf{r})$ is a vector pointing in the normal direction to the vortex plane. The contribution to the total vorticity along the edge C is measured by the dot product $[\nabla \times \mathbf{A}(\mathbf{r})] \cdot d\mathbf{S}$, which quantifies how aligned this vector is with the normal direction of the surface S. It is important to recall that the local vorticity at each point cancels out with the opposite vorticity at adjacent locations, leaving only the contribution along the "edge C" in the final result (see Fig. 2.11).

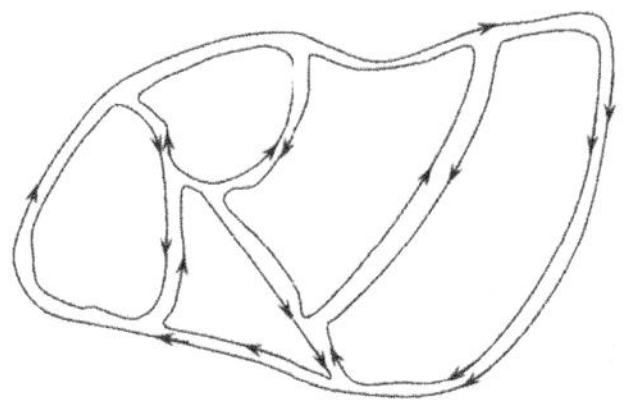

Fig. 2.11 The summation of vorticity. In regions where vortices are adjacent, their directions cancel each other out, leaving only the contributions along the outermost boundary

To summarize, we have skipped detailed derivations found in other texts and focused on how to interpret the theorems. It is essential to be familiar with the two expressions in Eq. (2.117), namely Gauss's theorem and Stokes's theorem, while keeping in mind the concept that "what happens within the region Ω is captured at its boundary $\partial\Omega$". It is important to remember the image in Fig. 2.10 and the meanings: "$\nabla \cdot \mathbf{A}(\mathbf{r})$ represents divergence" and "$\nabla \times \mathbf{A}(\mathbf{r})$ represents rotation". You should practice writing the equations by hand, while chanting, "summing up divergence gives the total inflow/outflow at the boundary" and "summing up rotation gives the total vorticity at the boundary". These concepts will often be used when converting "surface integrals to volume integrals" or "line integrals to surface integrals". Mastering these applications is the gateway to proficiency.

References

1. Riley KF, Hobson MP, Bence SJ (2006) Mathematical methods for physics and engineering, 3rd edn. Cambridge University Press, Cambridge. ISBN: 978-0521679718
2. Needham T (1999) Visual complex analysis. Clarendon Press, Oxford. ISBN: 978-0198534464
3. Kay DC (2011) Schaums outline tensor calculus. McGraw-Hill, New York. ISBN: 978-0071756037
4. Fleisch D (2011) A student's guide to vectors and tensors. Cambridge University Press, Cambridge. ISBN: 978-0521171908
5. VI Smirnov (1964) A course of higher mathematics. Pergamon, Oxford
6. Landau RH, Paez MJ, Bordeianu CC (2015) Computational physics: problem solving with python. Wiley-VCH, Weinheim. ISBN: 978-3527413157
7. Kompaneyets AS (2012) Theoretical physics, 2nd edn. Dover, Garden City. ISBN: 978-0486609720
8. Morse PM, Feshbach H (1953) Methods of theoretical physics. McGraw-Hill, New York. ISBN: 978-0070433168

Chapter 3
Essentials of Electromagnetism

As explained in Fig. 1.2, the ultimate goal of this book is to explain the origin of the '*coupling between spin and magnetic fields*'. This derivation will be discussed in Chap. 7 through relativistic quantum mechanics. In this theory, the electromagnetic field appears naturally. However, without describing the '*basic mathematical framework of electromagnetism*, the explanation in Chap. 7 will seem abrupt. With the foundational knowledge from Chap. 2, the concepts required to understand this are *almost within reach*. This chapter serves to provide the necessary fundamentals of electromagnetism for the logical developments in subsequent chapters. First, as an application of the field analysis introduced in the latter part of the previous chapter, we will formulate several concepts. We will then cover the differential laws for electrostatic fields and magnetostatic fields. Next, we will discuss electromagnetic induction and explore how the differential laws for electrostatic and magnetostatic fields are modified when dynamics are considered.

Repeatedly Learning Electromagnetism
Although electromagnetism is a subject typically studied in introductory courses, it remains profound and challenging, no matter how many times one revisits it. Achieving a truly deep understanding is elusive. That said, it is best to avoid being overly ambitious at the start. This chapter aims to provide a 'Level 1 overview' of the mathematical framework of electromagnetism. Beyond this, further levels can be envisioned: 'Level 2: Electromagnetism in media', 'Level 3: Electrodynamics based on Maxwell's equations', 'Level 4: Deep considerations on the significance of electric and magnetic flux density'. This multi-level approach offers a clearer roadmap (Sect. 9.3.4). In this book, we will focus on the laws governing electric fields **E** and magnetic fields **B** in a vacuum, leading to the derivation of **Maxwell's equations**. We will

(continued)

R. Maezono, *Essentials for Deeper Understanding of Quantum Computing*,
https://doi.org/10.1007/978-981-96-5646-2_3

also touch briefly on topics from 'Level 3', which covers electrodynamics rooted in Maxwell's equations. Although not discussed in this book, 'Level 3' includes topics such as how electric and magnetic fields transform into each other under Lorentz transformations of special relativity.

3.1 Applications of Field Analysis

3.1.1 Some Formulas for Inverse-Square Interactions

Using Gauss's theorem described in Sect. 2.5.8, we derive important equations that frequently appear in later discussions.

First, without requiring any special techniques, we can derive:

$$\nabla\left(\frac{1}{r}\right) = -\frac{\mathbf{r}}{r^3} = -\frac{\mathbf{e}_r}{r^2}, \quad \left(\mathbf{e}_r = \frac{\mathbf{r}}{r}\right). \tag{3.1}$$

Recalling that

$$r = \sqrt{x^2 + y^2 + z^2} = (x_l x_l)^{1/2},$$

this can be derived by performing the differential calculation:

$$\left[\nabla\left(\frac{1}{r}\right)\right]_j = \partial_j\left((x_l x_l)^{-1/2}\right) \tag{3.2}$$

(see Sect. 9.3.1).

Next, differentiating this again and calculating directly yields:

$$\nabla^2\left(\frac{1}{r}\right) = 0, \quad \ldots \quad (r \neq 0). \tag{3.3}$$

However, since $r = 0$ makes $1/r$ diverge, the differentiation process used to derive the above equation is not valid at $r = 0$. Thus, $r = 0$ is excluded from the domain of application.

For $r = 0$, we consider a closed surface S surrounding the singularity and apply Gauss's theorem to a volume integral inside:

$$\int_V \nabla^2\left(\frac{1}{r}\right) dV = \int_S \nabla\left(\frac{1}{r}\right) \cdot d\mathbf{S} = -\int_S \frac{\mathbf{e}_r}{r^2} \cdot d\mathbf{S}.$$

In the last equation, we substituted Eq. (3.1). Since the integrand is 0 away from the origin, we can shrink the closed surface S around the origin. If we take the closed surface S as a unit sphere, the unit vector $\mathbf{e}_r$ extending from the origin and the surface element $d\mathbf{S}$ point in the same direction. The magnitude of $d\mathbf{S}$ can be written as $r^2 d\Omega$.[1] Thus, $\mathbf{e}_r \cdot d\mathbf{S} = r^2 d\Omega$, so:[2]

$$\int_V \nabla^2 \left(\frac{1}{r}\right) dV = -\int_S \frac{\mathbf{e}_r}{r^2} \cdot d\mathbf{S} = -\int_S \frac{1}{r^2} \cdot r^2 d\Omega = -4\pi.$$

Revisiting the expression:

$$\int_V \nabla^2 \left(\frac{1}{r}\right) dV = -4\pi, \quad \nabla^2 \left(\frac{1}{r}\right) = 0 \quad \ldots \quad (r \neq 0),$$

this situation can be summarized using the delta function as:

$$\nabla^2 \left(\frac{1}{r}\right) = -4\pi\delta(x)\delta(y)\delta(z) = -4\pi\delta^3(\mathbf{r}). \tag{3.4}$$

3.1.2 Particle Interactions, Concept of Fields

The gravity we experience from birth, known as "things falling downward", is, in fact, the result of "the gravitational attraction between the Earth's mass M and an object's mass m", as taught in high school [1].[3] When the positions of the Earth's mass M and the object of mass m are $\mathbf{R}_j$ and $\mathbf{r}_i$, respectively, the gravitational force between the two bodies is expressed as:

$$\mathbf{F}\left(\mathbf{r}_i, \mathbf{R}_j\right) = -G \cdot \frac{m \cdot M}{|\mathbf{r}_i - \mathbf{R}_j|^2} \cdot \mathbf{e}_{[\mathbf{r}_i - \mathbf{R}_j]},$$

where G is the gravitational constant, and $\mathbf{e}_{[\mathbf{r}_i - \mathbf{R}_j]}$ is a unit vector pointing along the direction between the two masses. Note that both sides of the equation are vector

---

[1] $d\Omega = \sin\theta \cdot d\theta \cdot d\phi$, where $\theta = [0, \pi]$ and $\phi = [0, 2\pi]$ are the integration limits.

[2] 4π is the surface area of a unit sphere.

[3] However, when explaining to most university students that "all masses exert a weak gravitational pull on each other, and we feel gravity because Earth's M is significantly large", many respond with, "I heard this for the first time". Most people mistakenly believe, "Newton discovered that an apple falls from a tree". The wonder of Newton's discovery that the apple attracts the Earth just as the Earth attracts the apple, and that even a faint gravitational force exists between Newton and the apple, is often lost. Unless high school teachers explain it eloquently, many students misunderstand gravity as simply "the familiar downward force called gravitation".

quantities. Since gravity is an attractive force, the force **F** points in the opposite direction to $\mathbf{e}_r$, hence the negative sign.

Similarly, the **Coulomb electric force** is expressed as:

$$\mathbf{F}\left(\mathbf{r}_i, \mathbf{R}_j\right) = A \cdot \frac{q \cdot Q}{\left|\mathbf{r}_i - \mathbf{R}_j\right|^2} \cdot \mathbf{e}_{[\mathbf{r}_i - \mathbf{R}_j]},$$

as we learned this in high school [1]. Here, q and Q are electric charges, and the constant A is given by $A = 1/4\pi\varepsilon_0$.[4]

Taking the example of gravitation, if there are N masses $\left\{m_j, \mathbf{r}_j\right\}_{j=1}^N$, they mutually interact through gravitational forces. It is easy to imagine that if we attempt to describe this **many-body interaction system** by counting all pairs of interactions, the process becomes unmanageable. To address this, a more effective method is adopted: the **field concept**. For the multi-particle system $\left\{m_j, \mathbf{r}_j\right\}_{j=1}^N$, if we can determine what force a **test particle** m experiences when placed at a position **r**, then we can effectively describe the situation created by the N-particle system. The force felt by the test particle can be written as a function of position **r**:

$$\mathbf{F}(\mathbf{r}) = -G \sum_j \frac{m \cdot m_j}{\left|\mathbf{r} - \mathbf{r}_j\right|^2} \cdot \mathbf{e}_{[\mathbf{r} - \mathbf{r}_j]} = m \cdot \mathbf{E}(\mathbf{r}).$$

This function $\mathbf{E}(\mathbf{r})$ describes the **field** generated by the multi-particle system. The reason for factoring out m is to define $\mathbf{E}(\mathbf{r})$ as the field experienced per unit mass, reflecting a more fundamental description.

In the field concept, the **force field** $\mathbf{E}(\mathbf{r})$ provided by the multi-particle system describes the force experienced by a mass m placed at position **r**. This way of understanding corresponds to our traditional, experiential interpretation of gravity. We generally perceive our environment as if we are living in a "field where gravity is present", without consciously considering interacting partners like the Earth's mass M, as discovered by Newton. Now, by extending this concept, we interpret mass m as a 'coupling constant with the gravitational field'. A "heavier object with greater mass" can be understood as a "particle with a larger coupling constant to the gravitational field".[5]

Coulomb interaction systems are similarly expressed as:

$$\mathbf{F}(\mathbf{r}) = A \sum_j \frac{q \cdot q_j}{\left|\mathbf{r} - \mathbf{r}_j\right|^2} \cdot \mathbf{e}_{[\mathbf{r} - \mathbf{r}_j]} = q \cdot \mathbf{E}(\mathbf{r}). \tag{3.5}$$

---

[4] Refer to the footnote of Eq. (3.22) for the reason why the factor $1/4\pi$ is introduced.

[5] For example, instead of the basic idea that "a bag is heavy," the reader can now imagine that the bag "experiences a stronger force because of its stronger coupling to the gravitational field $\mathbf{E}(\mathbf{r})$".

This can be understood as "an environmental field $\mathbf{E}(\mathbf{r})$ (called the **electric field** in this case) created by charges other than the test particle, producing a force when the test particle is placed in this field". Here, the charge q can be interpreted as "a coupling constant with the electric field".[6]

3.1.3 Integral Representation Using Infinitesimal Contributions

The gravitational field

$$\mathbf{E}(\mathbf{r}) = -G\sum_j \frac{m_j}{\left|\mathbf{r}-\mathbf{r}_j\right|^2} \cdot \mathbf{e}_{[\mathbf{r}-\mathbf{r}_j]}$$

represents the force field experienced by a test particle at position $\mathbf{r}$ due to masses at each point j. This contribution can be expressed as

$$\Delta\mathbf{E}_j(\mathbf{r}) = -G\frac{m_j}{\left|\mathbf{r}-\mathbf{r}_j\right|^2} \cdot \mathbf{e}_{[\mathbf{r}-\mathbf{r}_j]}.$$

The total field is then

$$\mathbf{E}(\mathbf{r}) = \sum_j \Delta\mathbf{E}_j(\mathbf{r}),$$

as that summed over each contribution. Now, consider a continuous distribution of mass with density $\rho(\mathbf{r})$ instead of discrete masses. As shown in Fig. 3.1, a test particle at position $\mathbf{r}$ experiences an infinitesimal contribution from a small mass $\rho(\mathbf{r_j}) \cdot dV_j$ at position $\mathbf{r_j}$, given by

$$\delta\mathbf{E}_j(\mathbf{r}) = -G\frac{\rho(\mathbf{r}_j) \cdot dV_j}{\left|\mathbf{r}-\mathbf{r}_j\right|^2} \cdot \mathbf{e}_{[\mathbf{r}-\mathbf{r}_j]}.$$

[6] In the case of gravity, the "mass as a coupling constant" corresponds to the concept of "gravitational mass", which is contrasted with inertial mass [2, 3]. Regardless of the presence of gravity, we can sense "the mass of an object" through interactions like collisions, which aligns with the concept of inertial mass. Intuitively, the primary notion is "the mass that exists there", with gravitational mass being secondary. On the other hand, with electric charge, "the charge itself" lacks intuitive appeal; we recognize the presence of charge through electric forces. Therefore, defining it as a coupling constant (similar to gravitational mass) is more fundamental. This lack of intuitive grasp for charge complicates "the system of units in electromagnetism" [4, 5].

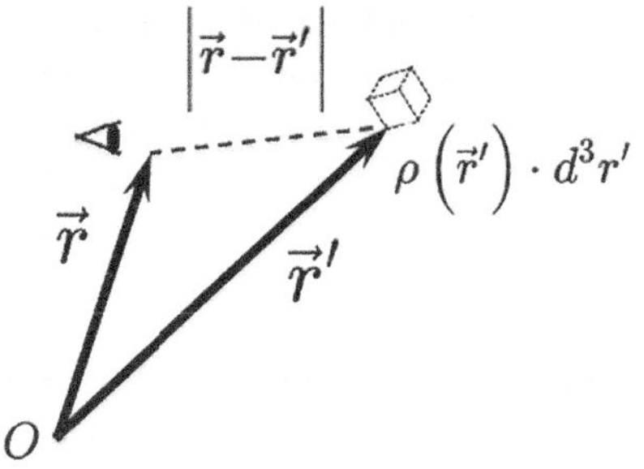

Fig. 3.1 The observation point **r**, where the test particle is placed, receives an infinitesimal contribution from the infinitesimal region at position $\mathbf{r}'$. By summing these contributions over "all $\mathbf{r}'$", the force field is evaluated

Summing these contributions, we have

$$\mathbf{E}(\mathbf{r}) = \sum_j \delta\mathbf{E}_j(\mathbf{r}) = -G\sum_j dV_j\,\frac{\rho\left(\mathbf{r}_j\right)}{\left|\mathbf{r}-\mathbf{r}_j\right|^2}\cdot\mathbf{e}_{[\mathbf{r}-\mathbf{r}_j]}.$$

Taking the continuous limit , the sum over j becomes an integral:[7]

$$\sum_j dV_j \to \int dV' = \int d^3r'.$$

Thus, the gravitational field becomes

$$\mathbf{E}(\mathbf{r}) = -G\int\frac{\rho\left(\mathbf{r}'\right)}{\left|\mathbf{r}-\mathbf{r}'\right|^2}\cdot\mathbf{e}_{[\mathbf{r}-\mathbf{r}']}\cdot d^3r'.$$

Pay close attention to the distinction between "the argument $\mathbf{r}$ of $\mathbf{E}(\mathbf{r})$" and "the integration variable $\mathbf{r}'$".[8]

For the electric field due to Coulomb forces, if the charge density at position $\mathbf{r}'$ is $\rho\left(\mathbf{r}'\right)$, the same reasoning leads to

$$\mathbf{E}(\mathbf{r}) = A\sum_j\frac{q_j}{\left|\mathbf{r}-\mathbf{r}_j\right|^2}\cdot\mathbf{e}_{[\mathbf{r}-\mathbf{r}_j]} \quad\to\quad A\int d^3r'\,\frac{\rho\left(\mathbf{r}'\right)}{\left|\mathbf{r}-\mathbf{r}'\right|^2}\cdot\mathbf{e}_{[\mathbf{r}-\mathbf{r}']}. \quad (3.6)$$

Please make sure to practice thoroughly so that you can swiftly write down such continuous transition expressions at any time.

[7] The notation d^3r is an abbreviation for $d^3r = dx\cdot dy\cdot dz$, which represents "an infinitesimal volume element in three-dimensional space.

[8] It is strongly recommended to write out this integral expression several times to solidify your understanding. Even at top university levels, I found some students struggle with this integral representation, not due to difficulty but due to a lack of careful explanation and comprehension.

3.1.4 *Gradient Flow and Potential*

The density $\rho(\mathbf{r})$ is a scalar field. The gradient $\nabla\rho(\mathbf{r})$ represents "the direction in which the density changes the most", as discussed in Sect. 2.5.5. If there is a gradient in density, a flow is established from regions of higher density to lower density. Since this flow moves in the direction of the steepest gradient, it can be expressed as:[9]

$$\mathbf{j}(\mathbf{r}) = -D \cdot \nabla\rho(\mathbf{r}).$$

The proportional constant D is sometimes called Einstein's coefficient . The direction of $\nabla\rho(\mathbf{r})$ points from regions of lower density to higher density. By introducing a negative sign, the direction is set to flow "from regions of higher density to lower density".

Behind the density gradient flow lies a "difference in density levels" that induces the flow, connected through the gradient operator. Concentrations or differences in levels are scalar fields, and a vector field representing the flow is formed according to these scalar variations. Many examples, such as atmospheric pressure and wind, water pressure and flow, and temperature and heat flow, can be understood through this relationship. If a scalar field $\phi(\mathbf{r})$ is given, a vector field $\mathbf{v}(\mathbf{r})$ can be constructed as $\mathbf{v}(\mathbf{r}) = -\nabla\phi(\mathbf{r})$. Thus, a scalar field can be seen as "a compressed form of information for the vector field." A scalar field with this role is called a **potential field**. The negative sign indicates that the flow moves *from regions of higher potential to regions of lower potential.*

3.1.5 *Irrotationality of Potential Flow*

As differential quantities of vector fields, the terms $\nabla \cdot \mathbf{A}(\mathbf{r})$ and $\nabla \times \mathbf{A}(\mathbf{r})$ were introduced (Sect. 2.5.6). Referring to the result only, **Helmholtz's theorem**[10] states that a vector field is uniquely determined by specifying $\nabla \cdot \mathbf{A}(\mathbf{r})$ and $\nabla \times \mathbf{A}(\mathbf{r})$.[11] Considering concepts like "a vector field without circulation" or "a vector field without divergence" can initially feel abrupt to beginners. However, due to Helmholtz's theorem, the two primary types of vector fields under analysis are the "irrotational field" and the "divergence-free field." Furthermore, as will be

---

9 This is known as Fick's Law , named after Adolf Eugen Fick (1829.09–1901.08), a German physicist and physiologist.

10 Hermann Ludwig Ferdinand von Helmholtz, 1821–1894, Germany.

11 The derivation of Helmholtz's theorem requires abstract formalism, which can be found, for instance, in [6, 7]. Nevertheless, the idea that a vector field is defined by the degree of its "circulation" and "divergence" aligns well with intuition.

discussed later, the concept of a **potential** is fundamentally tied to irrotationality, making it crucial to understand this concept thoroughly.

For a vector field that can be expressed as a gradient flow, $\mathbf{A} = -\nabla\varphi$, we can demonstrate that:

$$[\mathbf{A} = -\nabla\varphi] \quad \rightarrow \quad [\nabla \times \mathbf{A} = 0], \tag{3.7}$$

stating that gradient flow has no circulation. Writing $\mathbf{A} = -\nabla\varphi$ in component form, we have:

$$A_j = -\partial_j \varphi.$$

Similarly, expressing $\nabla \times \mathbf{A}$ in component form and substituting the above gives:

$$[\nabla \times \mathbf{A}]_j = \varepsilon_{jlm} \cdot \partial_l A_m = -\varepsilon_{jlm} \cdot \partial_l \partial_m \varphi.$$

Recalling Eq. (2.26), ε_{jlm} is antisymmetric with respect to (l, m), while $\partial_l \partial_m$ is symmetric. Therefore, this expression equals zero, which proves Eq. (3.7). The converse of Eq. (3.7),

$$[\nabla \times \mathbf{A} = 0] \rightarrow [\mathbf{A} = -\nabla\varphi],$$

can be mathematically proven and is known as *Poincare's Lemma*.[12]

3.1.6 Conservative Force Fields and Potentials

A force field $\mathbf{F}(\mathbf{r})$ can be visualized as a vector representing wind direction at each point in space. It is natural to think of a **potential field** $U(\mathbf{r})$, analogous to the "pressure behind the wind direction", such that:

$$\mathbf{F}(\mathbf{r}) = -\nabla U(\mathbf{r}). \tag{3.8}$$

For a force field expressed in this way as a gradient field, it follows from Eq. (3.7) that the field is irrotational:

$$\nabla \times \mathbf{F} = 0. \tag{3.9}$$

[12] This also requires advanced abstract formalism, which can be found, for instance, in [7].

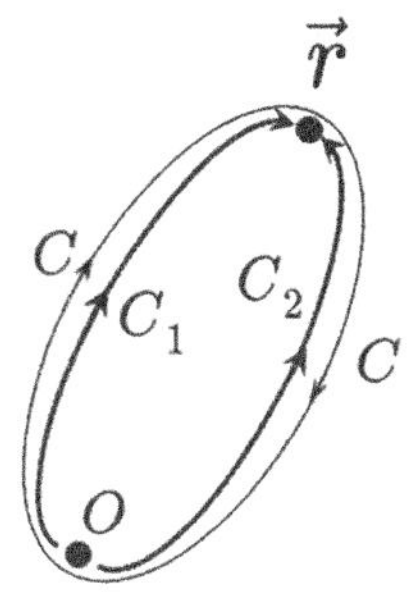

Fig. 3.2 On the closed path C, place the origin O and the position $\mathbf{r}$, then divide it into segments: "$C_1 : 0 \to \mathbf{r}$ (via A)" and "$C_2 : 0 \to \mathbf{r}$ (via B)"

Integrating this over a closed surface S,

$$0 = \int_S d\mathbf{S} \cdot (\nabla \times \mathbf{F}) = \oint_C d\mathbf{r} \cdot \mathbf{F},$$

we apply Stokes' theorem to express it as a line integral over a closed curve C. On the curve C, place the origin O and a position $\mathbf{r}$. Dividing the curve into two paths: "$C_1 : 0 \to \mathbf{r}$ (via A)" and "$C_2 : 0 \to \mathbf{r}$ (via B)" (Fig. 3.2). Thus,

$$0 = \oint_C d\mathbf{r} \cdot \mathbf{F} = \int_{C_1}^{\mathbf{r}} d\mathbf{r} \cdot \mathbf{F} - \int_{C_2}^{\mathbf{r}} d\mathbf{r} \cdot \mathbf{F}, \tag{3.10}$$

$$\therefore \quad \int_{C_1}^{\mathbf{r}} d\mathbf{r} \cdot \mathbf{F} = \int_{C_2}^{\mathbf{r}} d\mathbf{r} \cdot \mathbf{F} \tag{3.11}$$

is obtained. This integrand corresponds to the concept learned in high school, namely "(work) = (force) × (distance)". This quantity can be expressed as:[13]

$$\frac{d\mathbf{r}}{dt} \cdot \mathbf{F} = -\frac{d\mathbf{r}}{dt} \cdot \nabla U(\mathbf{r}) = -\frac{dr_l}{dt} \cdot \frac{\partial U(\mathbf{r})}{\partial r_l} = -\frac{dU(\mathbf{r})}{dt},$$
$$\therefore \quad d\mathbf{r} \cdot \mathbf{F} = -dU.$$

Thus,

$$\int^{\mathbf{r}(t)} d\mathbf{r} \cdot \mathbf{F}(\mathbf{r}) = -\int^{\mathbf{r}(t)} dU = -U(\mathbf{r}).$$

This establishes the connection to the potential. Since work has the dimension of energy, the potential $U(\mathbf{r})$ also has the dimension of energy. Therefore, the equation

[13] Recall Eq. (2.97).

derived from the fact that the vector field $\mathbf{F}(\mathbf{r})$ is irrotational, given by (3.11), becomes:

$$\int_{C_1}^{\mathbf{r}} d\mathbf{r} \cdot \mathbf{F} = \int_{C_2}^{\mathbf{r}} d\mathbf{r} \cdot \mathbf{F} = -U(\mathbf{r}).$$

The integral on the left-hand side corresponds to the "sum of work along a given path", which generally depends on the choice of the path. However, in the case of a **gradient force** given by Eq. (3.8), the rotation-free condition of Eq. (3.9) implies that the work integral does not depend on the path but is determined solely by the potential value $U(\mathbf{r})$ at the endpoint $\mathbf{r}$. This type of force field is called a **conservative force field** . At this point, it is essential to associate the concepts of "**path-independent line integral/potential definable/rotation-free/ gradient flow**" as mutually connected ideas.

3.1.7 Differential Expression of Particle Conservation Law

The flow of particles is represented as a vector field $\mathbf{j}(\mathbf{r})$. If the flow at a point $\mathbf{r}$ changes over time, it is expressed with time dependence as $\mathbf{j}(\mathbf{r}, t)$. Let the particle density be $\rho(\mathbf{r}, t)$. This density also changes with time, reflecting its time dependence. Now, consider a closed region V enclosed by an arbitrary closed surface S. Then,

$$\int_V dV \cdot \rho(\mathbf{r}, t),$$

represents the total number of particles within V.

Recalling Fig. 2.10a,

$$\int_S d\mathbf{S} \cdot \mathbf{j}(\mathbf{r}, t)$$

is the total particle flux $\mathbf{j}(\mathbf{r}, t)$ flowing out through the closed surface S. Since this must equal the decrease of particles within V,

$$-\frac{\partial}{\partial t}\left[\int_V dV \cdot \rho(\mathbf{r}, t)\right] = \int_S d\mathbf{S} \cdot \mathbf{j}(\mathbf{r}, t) \tag{3.12}$$

can be established. This is the integral expression for particle number conservation.

The right-hand side can be transformed by Gauss's integral theorem (Sect. 2.5.8) as

$$\int_S d\mathbf{S} \cdot \mathbf{j}(\mathbf{r}, t) = \int_V dV \cdot [\nabla \cdot \mathbf{j}(\mathbf{r}, t)].$$

Thus, Eq. (3.12) becomes

$$\int_V dV \cdot [\nabla \cdot \mathbf{j}(\mathbf{r}, t)] + \frac{\partial}{\partial t}\left[\int_V dV \cdot \rho(\mathbf{r}, t)\right] = 0,$$

$$\therefore \quad \int_V dV \cdot \left[\nabla \cdot \mathbf{j}(\mathbf{r}, t) + \frac{\partial \rho(\mathbf{r}, t)}{\partial t}\right] = 0,$$

expressed as a volume integral. Since this equation holds for any V, it follows that at any position $\mathbf{r}$, the integrand satisfies the *local law*

$$\nabla \cdot \mathbf{j}(\mathbf{r}, t) + \frac{\partial \rho(\mathbf{r}, t)}{\partial t} = 0. \tag{3.13}$$

This is the **local expression for particle number conservation.**

In a steady state with no time variation, the particle density satisfies

$$\frac{\partial \rho(\mathbf{r}, t)}{\partial t} = 0.$$

To satisfy Eq. (3.13), the particle flux must also respond accordingly, yielding

$$\nabla \cdot \mathbf{j}(\mathbf{r}, t) = 0. \tag{3.14}$$

This describes the fact that there are no sources or sinks in the particle flux, which is the mathematical expression for steady flow. This will later be used in the formulation of static magnetic fields.

3.2 Electrostatic and Magnetostatic Fields

According to Helmholtz's theorem (Sect. 3.1.5), a vector field $\mathbf{A}(\mathbf{r})$ can be uniquely determined if its divergence, $\nabla \cdot \mathbf{A}(\mathbf{r})$, and its curl, $\nabla \times \mathbf{A}(\mathbf{r})$, are specified. Thus, for the force fields of electricity [electric field $\mathbf{E}(\mathbf{r})$] and magnetism [magnetic field

$\mathbf{B}(\mathbf{r})$],[14] we will proceed to derive the differential relations, such as $\nabla \cdot \mathbf{E}(\mathbf{r})$ and $\nabla \times \mathbf{E}(\mathbf{r})$, that describe these fields.

3.2.1 Coulomb's Law and Electrostatic Field

From Eq. (3.6), the expression for the "field of force related to electric force" is given as

$$\begin{aligned}\mathbf{E}(\mathbf{r}) &= A\int d^3r' \frac{\rho(\mathbf{r}')}{|\mathbf{r}-\mathbf{r}'|^2} \cdot \mathbf{e}_{[\mathbf{r}-\mathbf{r}']} \\ &= \frac{1}{4\pi\varepsilon_0}\int d^3r' \frac{\rho(\mathbf{r}')}{|\mathbf{r}-\mathbf{r}'|^2} \cdot \mathbf{e}_{[\mathbf{r}-\mathbf{r}']}. \end{aligned} \tag{3.15}$$

The above equation provides the mathematical expression of **Coulomb's law** regarding electric force. The field $\mathbf{E}(\mathbf{r})$ is referred to as the **electric field**. However, since we will later handle $\mathbf{E}(\mathbf{r})$ with time dependence, we specifically call the time-independent electric field the **electrostatic field**.

Here, we confirm the relationship:

$$\frac{1}{|\mathbf{r}-\mathbf{r}'|^2} \cdot \mathbf{e}_{[\mathbf{r}-\mathbf{r}']} = \frac{\mathbf{r}-\mathbf{r}'}{|\mathbf{r}-\mathbf{r}'|^3}.$$

Building on this, recalling Eq. (3.1),

$$\nabla\left(\frac{1}{r}\right) = -\frac{\mathbf{r}}{r^3},$$

we can express:

$$\frac{1}{|\mathbf{r}-\mathbf{r}'|^2} \cdot \mathbf{e}_{[\mathbf{r}-\mathbf{r}']} = \frac{\mathbf{r}-\mathbf{r}'}{|\mathbf{r}-\mathbf{r}'|^3} = -\nabla_r\left(\frac{1}{|\mathbf{r}-\mathbf{r}'|}\right).$$

Here, ∇_r emphasizes differentiation with respect to $\mathbf{r}$, treating $\mathbf{r}'$ as a constant vector during the operation.

Thus, the expression for the electric field can be written as:

$$\mathbf{E}(\mathbf{r}) = -\frac{1}{4\pi\varepsilon_0}\int \rho(\mathbf{r}')\,\nabla_r\left(\frac{1}{|\mathbf{r}-\mathbf{r}'|}\right) \cdot d^3r'. \tag{3.16}$$

[14] Strictly speaking, this refers to the magnetic flux density. For a detailed discussion on this point, see Sect. 9.3.4.

Here, ∇_r is the differential operator with respect to $\mathbf{r}$, allowing it to be taken outside the integral over $\mathbf{r}'$:

$$\mathbf{E}(\mathbf{r}) = -\nabla_r \left[\frac{1}{4\pi\varepsilon_0} \int \frac{\rho(\mathbf{r}')}{|\mathbf{r}-\mathbf{r}'|} \cdot d^3r' \right] =: -\nabla_r \phi(\mathbf{r}).$$

Here, similar to Eq. (3.8), $\mathbf{F}(\mathbf{r}) = -\nabla U(\mathbf{r})$, the electric field can be represented as the gradient of a scalar potential:

$$\mathbf{E}(\mathbf{r}) = -\nabla_r\, \phi(\mathbf{r}). \tag{3.17}$$

This scalar potential is given by:

$$\phi(\mathbf{r}) = \frac{1}{4\pi\varepsilon_0} \int \frac{\rho(\mathbf{r}')}{|\mathbf{r}-\mathbf{r}'|} \cdot d^3r', \tag{3.18}$$

referred to as the **electrostatic potential** . Note that for an inverse-square force field, the potential is an inverse-linear function. Henceforth, become familiar with the concept that "**an inverse-square force arises from an inverse-linear potential**".

3.2.2 *Differential Laws of Electrostatic Fields*

As mentioned at the beginning of this section, we aim to evaluate the differential laws $\nabla \cdot \mathbf{E}(\mathbf{r})$ and $\nabla \times \mathbf{E}(\mathbf{r})$ to determine their respective forms.[15] First, concerning the rotation, recall Eq. (3.7), which states that "the rotation of a gradient field is zero". Since the electric field is expressed as a gradient field in Eq. (3.17),

$$\nabla \times \mathbf{E}(\mathbf{r}) = 0. \tag{3.19}$$

Next, we evaluate the divergence $\nabla \cdot \mathbf{E}(\mathbf{r})$. Using Eq. (3.16), we have

$$\begin{aligned} \nabla \cdot \mathbf{E}(\mathbf{r}) &= -\frac{1}{4\pi\varepsilon_0} \int \nabla_r \cdot \nabla_r \left(\frac{1}{|\mathbf{r}-\mathbf{r}'|} \right) \rho(\mathbf{r}') \cdot d^3r' \\ &= -\frac{1}{4\pi\varepsilon_0} \int \nabla_r^2 \left(\frac{1}{|\mathbf{r}-\mathbf{r}'|} \right) \rho(\mathbf{r}') \cdot d^3r'. \end{aligned}$$

[15] By evaluating these quantities, we can determine the vector field as stated at the beginning of this section.

From the identity in Eq. (3.4), we know

$$\nabla_r^2 \left(\frac{1}{|\mathbf{r} - \mathbf{r}'|} \right) = -4\pi \delta^3 \left(\mathbf{r} - \mathbf{r}' \right), \tag{3.20}$$

so the expression becomes

$$\begin{aligned} \nabla \cdot \mathbf{E}(\mathbf{r}) &= -\frac{1}{4\pi\varepsilon_0} \int \nabla_r^2 \left(\frac{1}{|\mathbf{r} - \mathbf{r}'|} \right) \rho \left(\mathbf{r}' \right) \cdot d^3 r' \\ &= \frac{1}{\varepsilon_0} \int \delta^3 \left(\mathbf{r} - \mathbf{r}' \right) \rho \left(\mathbf{r}' \right) \cdot d^3 r' \\ &= \frac{1}{\varepsilon_0} \rho (\mathbf{r}) . \end{aligned} \tag{3.21}$$

Thus, we arrive at

$$\nabla \cdot \mathbf{E}(\mathbf{r}) = \frac{\rho(\mathbf{r})}{\varepsilon_0}. \tag{3.22}$$

This equation is known as **Gauss's Law**.

Reason for the 4π Factor in the Coefficient
The coefficient $1/4\pi\varepsilon_0$ appearing in Eq. (3.15) may initially seem peculiar, leading to the question, "Why use a factor of $1/4\pi$ instead of a generic constant like A?" The rationale lies in Eq. (3.20), where a factor of 4π naturally emerges. By including the $1/4\pi$ factor in advance in Eq. (3.15), the cancellation of terms between numerator and denominator in Eq. (3.21) is ensured, simplifying the final expression.

3.2.3 Biot-Savart Law and the Static Magnetic Field

The static magnetic field expression corresponding to Coulomb's law for the static electric field (3.15) is referred to as the **Biot-Savart Law**, and it is given[16] by

$$\mathbf{B}(\mathbf{r}) = \frac{\mu_0}{4\pi} \int d^3 r' \frac{\mathbf{j}\left(\mathbf{r}'\right)}{|\mathbf{r} - \mathbf{r}'|^2} \times \mathbf{e}_{\mathbf{r}-\mathbf{r}'}. \tag{3.23}$$

[16] The inclusion of the $1/4\pi$ factor in the coefficient follows the same logic as discussed in the column for Eq. (3.22).

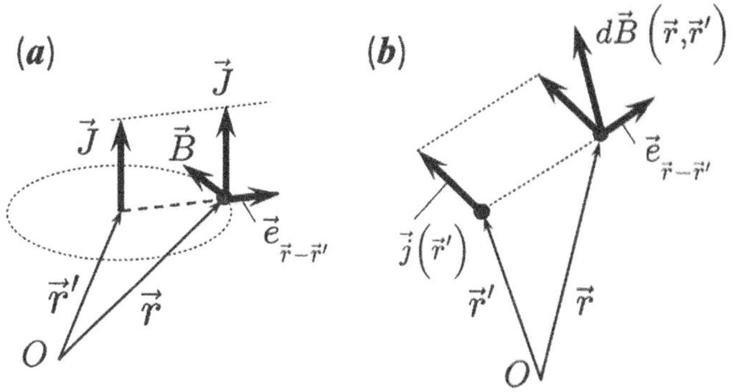

Fig. 3.3 Geometry of the Biot-Savart law. Panel (**a**) illustrates the magnetic field **B** generated by a steady current **J** directed vertically. This can be generalized by considering the contribution $d\mathbf{B}(\mathbf{r},\mathbf{r}')$ from a current element $\mathbf{j}(\mathbf{r}')$ at each location [Panel (**b**)]. Contributions from other locations, such as $\mathbf{j}(\mathbf{r}'')$ and $\mathbf{j}(\mathbf{r}''')$, are similarly accounted for

To fully understand this formula, we begin with a detailed explanation. Familiarity with this expression is important; eventually, one should be able to recite and write it down effortlessly.[17]

Recall from high school physics that a magnetic field **B** is generated around a current **j** following the right-hand rule [Fig. 3.3a]. This directional relationship can be expressed using the cross product as $\mathbf{B} \propto (\mathbf{J} \times \mathbf{e}_{\mathbf{r}-\mathbf{r}'})$. Similarly, as also learned in high school, the magnitude of the magnetic field decays inversely with the square of the distance from the current, leading to

$$\mathbf{B} = A \cdot \frac{\mathbf{J} \times \mathbf{e}_{\mathbf{r}-\mathbf{r}'}}{|\mathbf{r}-\mathbf{r}'|^2}. \tag{3.24}$$

The proportionality constant is conventionally written as $A = \mu_0/4\pi$ (Sect. 9.3.4).

Once you understand Eq. (3.24), recall the discussion in Sect. 3.1.3 about "summing the contributions from elements". At the observation point **r** in Fig. 3.3b, the contribution from each current element $\mathbf{j}(\mathbf{r}')$ is given by[18]

$$d\mathbf{B}(\mathbf{r},\mathbf{r}') = \frac{\mu_0}{4\pi} \cdot \frac{\mathbf{j}(\mathbf{r}') \times \mathbf{e}_{\mathbf{r}-\mathbf{r}'}}{|\mathbf{r}-\mathbf{r}'|^2}.$$

[17] Practicing the relationship between the directions of the magnetic field **B** and the current **j** by writing the cross-product formula on paper several times helps to internalize the "sense of the cross-product".

[18] The quantity $d\mathbf{B}(\mathbf{r},\mathbf{r}')$ is written to explicitly show its dependence on $(\mathbf{r},\mathbf{r}')$. Initially, one might simply copy this notation from a teacher's explanation, but it is important to eventually develop the ability to independently decide "how to assign such arguments" for clarity and precision. This dependence reflects the contribution "from the element at position $\mathbf{r}'$ to the observation point $\mathbf{r}$". Hence, the arguments are structured to represent this relationship clearly.

Summing these contributions over all element positions $\mathbf{r}'$ yields

$$\mathbf{B}(\mathbf{r}) = \int d^3r' \, d\mathbf{B}(\mathbf{r}, \mathbf{r}'),$$

which results in Eq. (3.23).

Once an understanding of the Biot-Savart law, as given by Eq. (3.23), has been established, the next step is to apply the formula

$$\frac{\mathbf{e}_{\mathbf{r}-\mathbf{r}'}}{|\mathbf{r}-\mathbf{r}'|^2} = -\nabla_r \left(\frac{1}{|\mathbf{r}-\mathbf{r}'|} \right)$$

to express the magnetic field as

$$\mathbf{B}(\mathbf{r}) = -\frac{\mu_0}{4\pi} \int d^3r' \cdot \mathbf{j}(\mathbf{r}') \times \nabla_r \left(\frac{1}{|\mathbf{r}-\mathbf{r}'|} \right). \tag{3.25}$$

3.2.4 Vector Potential

In the integrand of Eq. (3.25), the operator ∇_r acts as a differential operator with respect to $\mathbf{r}$. Therefore, when considering its action, $\mathbf{j}(\mathbf{r}')$ can be treated as a constant vector $\mathbf{j}$. For simplicity, let us denote $\phi(\mathbf{r}) = |\mathbf{r}-\mathbf{r}'|$, allowing us to derive the following relation:

$$\mathbf{j} \times \nabla_r \left(\frac{1}{\phi(\mathbf{r})} \right) = -\nabla_r \times \left(\frac{\mathbf{j}}{\phi(\mathbf{r})} \right), \tag{3.26}$$

as derived in detail in Sect. 9.3.2. Using this, Eq. (3.25) can be rewritten as:[19]

$$\begin{aligned} \mathbf{B}(\mathbf{r}) &= \frac{\mu_0}{4\pi} \int d^3r' \cdot \nabla_r \times \left(\frac{\mathbf{j}(\mathbf{r}')}{|\mathbf{r}-\mathbf{r}'|} \right) = \nabla_r \times \left[\frac{\mu_0}{4\pi} \int d^3r' \cdot \left(\frac{\mathbf{j}(\mathbf{r}')}{|\mathbf{r}-\mathbf{r}'|} \right) \right] \\ &= \nabla_r \times \mathbf{A}(\mathbf{r}). \end{aligned}$$

In this expression:

$$\mathbf{A}(\mathbf{r}) = \frac{\mu_0}{4\pi} \int d^3r' \cdot \left(\frac{\mathbf{j}(\mathbf{r}')}{|\mathbf{r}-\mathbf{r}'|} \right) \tag{3.27}$$

[19] Here, ∇_r is a differential operator with respect to $\mathbf{r}$, which allows it to be taken outside the integral over $\mathbf{r}'$.

is defined as the **vector potential**. Thus, the magnetic field $\mathbf{B}(\mathbf{r})$ can be expressed as the rotation of the vector potential $\mathbf{A}(\mathbf{r})$:

$$\mathbf{B}(\mathbf{r}) = \nabla_r \times \mathbf{A}(\mathbf{r}) .$$

This establishes the relationship between the magnetic field and the vector potential.

To summarize alongside the case of the electric field, we have:

$$\mathbf{E}(\mathbf{r}) = -\nabla_r \phi(\mathbf{r}), \quad \phi(\mathbf{r}) = \frac{1}{4\pi\varepsilon_0}\int d^3r' \cdot \frac{\rho(\mathbf{r}')}{|\mathbf{r}-\mathbf{r}'|}$$

$$\mathbf{B}(\mathbf{r}) = \nabla_r \times \mathbf{A}(\mathbf{r}), \quad \mathbf{A}(\mathbf{r}) = \frac{\mu_0}{4\pi}\int d^3r' \cdot \frac{\mathbf{j}(\mathbf{r}')}{|\mathbf{r}-\mathbf{r}'|}$$

Thus, the electric field is expressed as the gradient of a scalar field, and the magnetic field as the curl of a vector field. The scalar potential and vector potential are computed by aggregating the contributions from their respective sources—charge density and current density—with a weight factor of $1/\left|\mathbf{r}-\mathbf{r}'\right|$. These formulations exhibit a symmetric structure.

3.2.5 *Differential Rules for the Vector Potential*

The vector potential $\mathbf{A}(\mathbf{r})$, being a vector field, is characterized by its rotation $\nabla_r \times \mathbf{A}(\mathbf{r})$ and its divergence $\nabla_r \cdot \mathbf{A}(\mathbf{r})$.[20] Here, we calculate these differential properties. Firstly, for the rotation, by definition:

$$\nabla_r \times \mathbf{A}(\mathbf{r}) = \mathbf{B}(\mathbf{r}) .$$

Next, to evaluate the divergence $\nabla_r \cdot \mathbf{A}(\mathbf{r})$ for (3.27), we have:

$$\nabla_r \cdot \mathbf{A}(\mathbf{r}) = \frac{\mu_0}{4\pi}\int d^3r' \cdot \mathbf{j}(\mathbf{r}')\,\nabla_r\left(\frac{1}{|\mathbf{r}-\mathbf{r}'|}\right).$$

Using a technical substitution, we rewrite:

$$\nabla_r\left(\frac{1}{|\mathbf{r}-\mathbf{r}'|}\right) = -\nabla_{r'}\left(\frac{1}{|\mathbf{r}-\mathbf{r}'|}\right),$$

[20] As mentioned at the beginning of this section, a vector field can be determined once its rotation and divergence are known.

which allows us to switch from derivatives with respect to r to those with respect to r', yielding:

$$\nabla_r \cdot \mathbf{A}(\mathbf{r}) = -\frac{\mu_0}{4\pi}\int d^3r' \cdot \mathbf{j}(\mathbf{r}')\, \nabla_{r'}\left(\frac{1}{|\mathbf{r}-\mathbf{r}'|}\right).$$

For the integrand, we apply the identity:

$$\mathbf{j}(\mathbf{r}') \cdot \nabla_{r'}\left(\frac{1}{|\mathbf{r}-\mathbf{r}'|}\right) = \nabla_{r'}\left(\frac{\mathbf{j}(\mathbf{r}')}{|\mathbf{r}-\mathbf{r}'|}\right) - \frac{\nabla_{r'} \cdot \mathbf{j}(\mathbf{r}')}{|\mathbf{r}-\mathbf{r}'|}.$$

Here, since we are discussing static magnetic fields, the current density can be assumed to be steady, and from (3.14), we have:

$$\nabla_{r'} \cdot \mathbf{j}(\mathbf{r}') = 0.$$

Using this, the divergence of the vector potential simplifies to:

$$\nabla_r \cdot \mathbf{A}(\mathbf{r}) = -\frac{\mu_0}{4\pi}\int d^3r' \cdot \nabla_{r'}\left(\frac{\mathbf{j}(\mathbf{r}')}{|\mathbf{r}-\mathbf{r}'|}\right).$$

This volume integral can then be transformed using Gauss's theorem [(2.117)], giving:

$$\nabla_r \cdot \mathbf{A}(\mathbf{r}) = -\frac{\mu_0}{4\pi}\int d^3r' \cdot \nabla_{r'}\left(\frac{\mathbf{j}(\mathbf{r}')}{|\mathbf{r}-\mathbf{r}'|}\right) = -\frac{\mu_0}{4\pi}\int_S d\mathbf{S} \cdot \frac{\mathbf{j}(\mathbf{r}')}{|\mathbf{r}-\mathbf{r}'|}. \quad (3.28)$$

Here, let us consider the region where the volume integral in the above equation and the enclosing closed surface S are defined. The volume integral over d^3r' extends over the region where the current density $\mathbf{j}(\mathbf{r}')$ exists. Thus, the closed surface S represents the boundary enclosing this region of non-zero current density. Since the current $\mathbf{j}$ forms a closed loop, we can choose the closed surface S to be sufficiently far away so that no current crosses S. In such a case, the term $d\mathbf{S} \cdot \mathbf{j}$ becomes:

$$\nabla_r \cdot \mathbf{A}(\mathbf{r}) = -\frac{\mu_0}{4\pi}\int_S d\mathbf{S} \cdot \frac{\mathbf{j}(\mathbf{r}')}{|\mathbf{r}-\mathbf{r}'|} = 0.$$

In summary, the differential relations for the vector potential $\mathbf{A}(\mathbf{r})$ are:

$$\nabla_r \times \mathbf{A}(\mathbf{r}) = \mathbf{B}(\mathbf{r}), \quad \nabla_r \cdot \mathbf{A}(\mathbf{r}) = 0. \quad (3.29)$$

3.2.6 Differential Laws of Static Magnetic Fields

Next, we evaluate the differential laws for the magnetic field $\mathbf{B}(\mathbf{r})$, a vector field, focusing on its rotation $\nabla_r \times \mathbf{B}(\mathbf{r})$ and divergence $\nabla_r \cdot \mathbf{B}(\mathbf{r})$. Regarding the divergence, from the identity that "the divergence of a rotation is identically zero", we have:

$$\nabla_r \cdot \mathbf{B}(\mathbf{r}) = \nabla_r \cdot [\nabla_r \times \mathbf{A}(\mathbf{r})] = 0. \tag{3.30}$$

(See Sect. 9.3.2).

For the rotation, we evaluate:

$$\nabla_r \times \mathbf{B}(\mathbf{r}) = \nabla_r \times [\nabla_r \times \mathbf{A}(\mathbf{r})].$$

Using the identity for the double rotation of a vector field, we obtain:

$$\nabla_r \times \mathbf{B}(\mathbf{r}) = \nabla_r \times [\nabla_r \times \mathbf{A}(\mathbf{r})] = \nabla_r [\nabla_r \cdot \mathbf{A}(\mathbf{r})] - \nabla_r^2 (\mathbf{A}(\mathbf{r})), \tag{3.31}$$

(See Sect. 9.3.2). Furthermore, from (3.29), we know:

$$\nabla_r \cdot \mathbf{A}(\mathbf{r}) = 0.$$

Substituting (3.27) into the equation and continuing the evaluation, we find:

$$\nabla_r \times \mathbf{B}(\mathbf{r}) = -\nabla_r^2 (\mathbf{A}(\mathbf{r})) = -\frac{\mu_0}{4\pi} \int d^3r' \cdot \mathbf{j}(\mathbf{r}') \nabla_r^2 \left(\frac{1}{|\mathbf{r} - \mathbf{r}'|} \right).$$

Applying the formula:

$$\nabla_r^2 \left(\frac{1}{|\mathbf{r} - \mathbf{r}'|} \right) = -4\pi \delta^3 (\mathbf{r} - \mathbf{r}'),$$

we arrive at:[21]

$$\nabla_r \times \mathbf{B}(\mathbf{r}) = -\frac{\mu_0}{4\pi} \int d^3r' \cdot \mathbf{j}(\mathbf{r}') \left[-4\pi \delta^3 (\mathbf{r} - \mathbf{r}') \right] = \mu_0 \mathbf{j}(\mathbf{r}).$$

In summary, for the static magnetic field $\mathbf{B}(\mathbf{r})$, we have:

$$\nabla_r \cdot \mathbf{B}(\mathbf{r}) = 0, \quad \nabla_r \times \mathbf{B}(\mathbf{r}) = \mu_0 \mathbf{j}(\mathbf{r}).$$

[21] Here, the 4π factor in the denominator of the coefficient, preemptively extracted, cancels with the 4π factor in the delta function coefficient, as intended.

When we also include the case for the static electric field $\mathbf{E}(\mathbf{r})$, we get:

$$\nabla \times \mathbf{E}(\mathbf{r}) = 0, \quad \nabla \cdot \mathbf{E}(\mathbf{r}) = \frac{\rho(\mathbf{r})}{\varepsilon_0}.$$

These four equations form the governing equation set for stationary "static electromagnetic fields" in the absence of time variation.

3.3 Laws for Dynamic Electromagnetic Fields

In the previous section, we derived the governing equations for static electromagnetic fields. In this section, we will consider how the rotation and divergence of $\mathbf{B}(\mathbf{r}, t)$ and $\mathbf{E}(\mathbf{r}, t)$ behave when the fields exhibit time variations. To state the conclusion upfront, the governing equations for static electromagnetic fields undergo modifications specifically in $\nabla_r \times \mathbf{B}(\mathbf{r}, t)$ and $\nabla \times \mathbf{E}(\mathbf{r}, t)$. The former is influenced by the concept of displacement current, while the latter is derived from electromagnetic induction.

3.3.1 Electric Displacement and Magnetization

Focusing on the source-related equations among the governing equations for static electromagnetic fields, we can write them as follows:

$$\rho(\mathbf{r}) = \nabla \cdot \varepsilon_0 \mathbf{E}(\mathbf{r}), \quad \mathbf{j}(\mathbf{r}) = \nabla \times \frac{\mathbf{B}(\mathbf{r})}{\mu_0}.$$

It is convenient to introduce fields defined with coefficients as follows:

$$\varepsilon_0 \mathbf{E}(\mathbf{r}) =: \mathbf{D}(\mathbf{r}), \quad \frac{\mathbf{B}(\mathbf{r})}{\mu_0} =: \mathbf{H}(\mathbf{r}). \tag{3.32}$$

Using these definitions, the governing equations for static electromagnetic fields become:

$$\begin{aligned} \nabla \cdot \mathbf{B}(\mathbf{r}) = 0, \quad \nabla \times \mathbf{H}(\mathbf{r}) = \mathbf{j}(\mathbf{r}), \\ \nabla \times \mathbf{E}(\mathbf{r}) = 0, \quad \nabla \cdot \mathbf{D}(\mathbf{r}) = \rho(\mathbf{r}). \end{aligned} \tag{3.33}$$

Here, the introduced fields **D** and **H** are called **electric displacement** and **magnetic field** , respectively. In this context, the field previously referred to as the magnetic field, **B**, is now called the **magnetic flux density**.[22]

When considering the introduction of the electric displacement **D** and the magnetic field **H**, as explained in this book within the context of vacuum theory, they might simply seem like a "convenient redefinition" of the electric field **E** and the magnetic flux density **B**. However, in electromagnetic theory within media [8, 9], **D** and **H** are distinguished from **E** and **B**, and careful discussions on "which field induces what, and which field is generated by that" lead to an independent treatment of these four fields.

3.3.2 *Displacement Current*

Recall that "the divergence of a rotation is identically zero" (as derived in (3.30) in Sect. 9.3.2). Using this, taking the divergence of (3.33), $\mathbf{j}(\mathbf{r}) = \nabla \times \mathbf{H}(\mathbf{r})$, yields:

$$\nabla \cdot \mathbf{j}(\mathbf{r}) = \nabla \cdot [\nabla \times \mathbf{H}(\mathbf{r})] = 0. \tag{3.34}$$

On the other hand, the conservation of charge as described in (3.13) is expressed as:

$$\nabla \cdot \mathbf{j}(\mathbf{r}, t) = -\frac{\partial \rho(\mathbf{r}, t)}{\partial t}. \tag{3.35}$$

Note that the conservation of charge was derived from the outset assuming dynamic conditions, with the arguments $(\mathbf{r}, t)$. In contrast, the equations for static magnetic fields were developed without the presence of time t in the arguments. In the static case, where $\partial \rho(\mathbf{r}) / \partial t = 0$, (3.35) reduces to (3.34).

To extend (3.34) to dynamic cases, it should satisfy (3.35). Let us introduce an additional term $\mathbf{X}(\mathbf{r}, t)$ such that:

$$\mathbf{j}(\mathbf{r}, t) = \nabla \times \mathbf{H}(\mathbf{r}, t) + \mathbf{X}(\mathbf{r}, t). \tag{3.36}$$

Taking the divergence of $\mathbf{j}(\mathbf{r}, t)$, we have:

$$\nabla \cdot \mathbf{j}(\mathbf{r}, t) = \nabla \cdot [\nabla \times \mathbf{H}(\mathbf{r}, t) + \mathbf{X}(\mathbf{r}, t)] = \nabla \cdot \mathbf{X}(\mathbf{r}, t).$$

Comparing this result with (3.35), we find:

$$\nabla \cdot \mathbf{X}(\mathbf{r}, t) = -\frac{\partial \rho(\mathbf{r}, t)}{\partial t}.$$

[22] In contexts without ambiguity, **B** may still be referred to as the magnetic field (Sect. 9.3.4).

Thus, the additional term $\mathbf{X}(\mathbf{r}, t)$ must account for the time-dependent divergence of charge density.

From (3.33), we know:

$$\rho(\mathbf{r}) = \nabla \cdot \mathbf{D}(\mathbf{r}),$$

which holds at every moment in time. Thus, this relationship also extends to time-varying cases:[23]

$$\rho(\mathbf{r}, t) = \nabla \cdot \mathbf{D}(\mathbf{r}, t).$$

Using this result, we find:

$$\nabla \cdot \mathbf{X}(\mathbf{r}, t) = -\frac{\partial}{\partial t}\left[\nabla \cdot \mathbf{D}(\mathbf{r}, t)\right].$$

Since spatial and time derivatives commute, we can write:

$$\mathbf{X}(\mathbf{r}, t) = -\frac{\partial}{\partial t}\mathbf{D}(\mathbf{r}, t).$$

This additional term $\mathbf{X}(\mathbf{r}, t)$ represents the time derivative of the displacement field **D**.

Substituting into (3.36), we have:

$$\mathbf{j}(\mathbf{r}, t) = \nabla \times \mathbf{H}(\mathbf{r}, t) - \frac{\partial}{\partial t}\mathbf{D}(\mathbf{r}, t)$$

$$\therefore \quad \nabla \times \mathbf{H}(\mathbf{r}, t) = \mathbf{j}(\mathbf{r}, t) + \frac{\partial \mathbf{D}(\mathbf{r}, t)}{\partial t} \tag{3.37}$$

Thus, the differential law for static magnetic fields, $\nabla \times \mathbf{H}(\mathbf{r}) = \mathbf{j}(\mathbf{r})$, is extended and modified into the "differential law for dynamic magnetic fields" as given in (3.37).

The term $\partial \mathbf{D}(\mathbf{r}, t)/\partial t$ is an additional term contributing to the current in the presence of time-dependence. This is called the **displacement current**. For instance, a capacitor blocks direct current (DC) but allows current to flow when an alternating voltage is applied, a phenomenon explained by the displacement current.

[23] This equation was derived for the case of static electric fields. Thus, whether it remains valid without any additional corrective terms when time-dependent variables are introduced requires careful consideration. As a conclusion, it does indeed hold without modification, but many textbooks [4, 10] tend to omit explicit mention of the time-dependent arguments $(\mathbf{r}, t)$ in their discussions, and this aspect is often not addressed. Older references, such as [11, 12], provide limited commentary on this point, with brief annotations.

In (3.37), it is also important to note that the presence of time-dependence introduces a **coupling between electric and magnetic fields**.

3.3.3 Electromagnetic Induction

In the preceding section on displacement currents, we concluded that "the rotation $\nabla \times \mathbf{H}(\mathbf{r}, t)$ of the magnetic field is coupled to the time derivative of the electric field". In this section, we derive a complementary concept: "the rotation of the electric field is coupled to the time derivative of the magnetic field". This arises as a mathematical description of the experimental phenomenon known as electromagnetic induction.

Electromagnetic induction is a phenomenon familiar from middle school science, where "a strong movement of a magnet towards a coil connected to a voltmeter generates a voltage". The phenomenon is expressed as "the greater the time derivative in magnetic flux Φ, the larger the induced voltage V":

$$-\frac{\partial \Phi(t)}{\partial t} \propto V(t). \tag{3.38}$$

The negative sign indicates that "the induced voltage acts to oppose the change in magnetic flux". The magnetic flux Φ is the summation of the magnetic field passing through a surface S bounded by a closed loop C formed by the coil (Fig. 3.4):

$$\Phi(t) = \int_S \mathbf{B}(\mathbf{r}, t) \cdot d\mathbf{S}$$

Here, the dot product accounts for "how much the field directly pierces the surface".

The term "voltage V induced in a closed circuit" refers to the electromotive force, which is represented by an electric field $\mathbf{E}_{\mathrm{ex}}$ *distinct from the static electric field.*

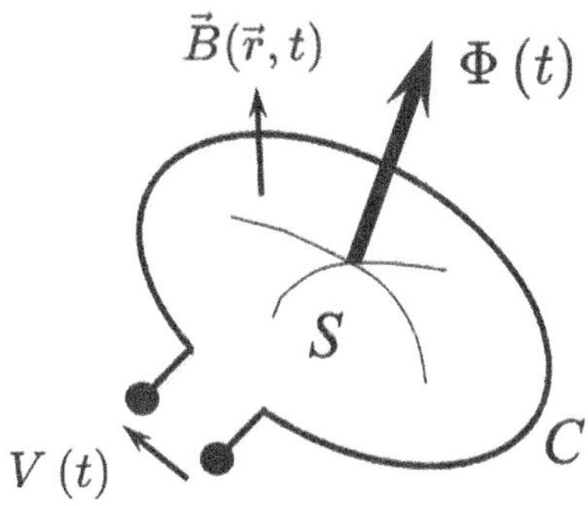

Fig. 3.4 The magnetic flux Φ represents the summation of the magnetic field passing through the surface S bounded by the closed loop C formed by the coil. The temporal rate of change of the magnetic flux is proportional to the potential difference $V(t)$ generated across the ends of the coil

This is expressed as (Sect. 9.3.3):

$$V(t) = \int_C \mathbf{E}_{ex} \cdot d\mathbf{l}.$$

Considering the dynamic electric field $\mathbf{E}(\mathbf{r}, t)$, which includes $\mathbf{E}_{ex}$, as:

$$\mathbf{E}(\mathbf{r}, t) = \mathbf{E}(\mathbf{r}) + \mathbf{E}_{ex}(\mathbf{r}, t),$$

the electromotive force becomes:

$$V(t) = \int_C \mathbf{E}(\mathbf{r}, t) \cdot d\mathbf{l}.$$

The law of electromagnetic induction, Eq. (3.38), then takes the form:

$$-\frac{\partial}{\partial t}\int_S \mathbf{B} \cdot d\mathbf{S} = \int_C \mathbf{E}(\mathbf{r}, t) \cdot d\mathbf{l}.$$

Applying Stokes' theorem, Eq. (2.117):

$$\int_C \mathbf{E}(\mathbf{r}, t) \cdot d\mathbf{l} = \int_S [\nabla \times \mathbf{E}(\mathbf{r}, t)] \cdot d\mathbf{S},$$

yields:

$$-\frac{\partial}{\partial t}\int_S \mathbf{B} \cdot d\mathbf{S} = \int_S [\nabla \times \mathbf{E}(\mathbf{r}, t)] \cdot d\mathbf{S}$$

$$\therefore \quad \int_S \left[\nabla \times \mathbf{E}(\mathbf{r}, t) + \frac{\partial \mathbf{B}(\mathbf{r}, t)}{\partial t}\right] \cdot d\mathbf{S} = 0.$$

Since this holds for any arbitrary surface S, the integrand must satisfy the identity:

$$\nabla \times \mathbf{E}(\mathbf{r}, t) + \frac{\partial \mathbf{B}(\mathbf{r}, t)}{\partial t} = 0. \tag{3.39}$$

This equation serves as the **differential expression for electromagnetic induction phenomena.**[24]

[24] To clarify, a "differential expression" describes a phenomenon as a quantitative relationship that must hold at any position and time.

3.4 Maxwell's Equations and Electromagnetic Waves

The laws governing electric and magnetic fields discussed thus far are consolidated into a series of equations written in the framework of vector calculus, known as Maxwell's equations (Sect. 3.4.1). These results will be utilized in the latter part of this text (Fig. 1.2). The section on electromagnetic waves (Sect. 3.4.2) is included to bridge the connections indicated by the dashed lines in Fig. 1.2. It serves as a foundational element when discussing the optical-mechanical analogy (Sect. 5.1.1), providing a reference for better understanding.

3.4.1 *Maxwell's Equations*

Rewriting the differential laws governing electric and magnetic fields for dynamic scenarios, we obtain:

$$\nabla \cdot \mathbf{B}(\mathbf{r}, t) = 0 , \quad \nabla \times \mathbf{H}(\mathbf{r}, t) = \mathbf{j}(\mathbf{r}, t) + \underline{\frac{\partial \mathbf{D}(\mathbf{r}, t)}{\partial t}}$$

$$\nabla \times \mathbf{E}(\mathbf{r}, t) = \underline{-\frac{\partial \mathbf{B}(\mathbf{r}, t)}{\partial t}} , \quad \nabla \cdot \mathbf{D}(\mathbf{r}, t) = \rho(\mathbf{r}, t)$$

These are known as **Maxwell's equations**. Note that allowing for time-dependence introduces coupling between the electric and magnetic fields through the underlined terms.

Dynamic electromagnetic fields relate to potentials as follows. Substituting $\mathbf{B} = \nabla \times \mathbf{A}$ into the electromagnetic induction equation (3.39) gives:

$$\nabla \times \left(\mathbf{E}(\mathbf{r}, t) + \frac{\partial \mathbf{A}(\mathbf{r}, t)}{\partial t} \right) = 0.$$

From Poincare's lemma (Sect. 3.1.5), a rotation-free field can be expressed as the gradient of a scalar potential:

$$\mathbf{E}(\mathbf{r}, t) + \frac{\partial \mathbf{A}(\mathbf{r}, t)}{\partial t} = -\nabla \phi(\mathbf{r}, t) .$$

Here, $\phi(\mathbf{r}, t)$ is the scalar potential. In the absence of time-dependence, the term $\partial \mathbf{A}(\mathbf{r}, t)/\partial t$ vanishes, reducing this equation to the static case, reproducing the potentials $\phi(\mathbf{r})$ as in (3.17) and (3.18).

In summary, combining the vector potential for the magnetic field, the relationship between the potentials $(\phi, \mathbf{A})$ and the electromagnetic fields $(\mathbf{E}, \mathbf{B})$ is given as:

$$\mathbf{B}(\mathbf{r}, t) = \nabla \times \mathbf{A}(\mathbf{r}, t), \quad \mathbf{E}(\mathbf{r}, t) = -\nabla\phi(\mathbf{r}, t) - \frac{\partial \mathbf{A}(\mathbf{r}, t)}{\partial t}. \tag{3.40}$$

3.4.2 Electromagnetic Waves

The time evolution of the electric and magnetic fields derived from Maxwell's equations can be expressed as:

$$\begin{aligned} \frac{\partial \mathbf{D}(\mathbf{r}, t)}{\partial t} &= \nabla \times \mathbf{H}(\mathbf{r}, t) - \mathbf{j}(\mathbf{r}, t), \\ \frac{\partial \mathbf{B}(\mathbf{r}, t)}{\partial t} &= -\nabla \times \mathbf{E}(\mathbf{r}, t). \end{aligned}$$

Restricting to the vacuum, where $\mathbf{j}(\mathbf{r}, t) = 0$, and substituting (3.32) into these equations to express them solely in terms of the electric field $\mathbf{E}$ and magnetic flux density $\mathbf{B}$, we have:

$$\begin{aligned} \varepsilon_0\mu_0 \cdot \frac{\partial \mathbf{E}(\mathbf{r}, t)}{\partial t} &= \nabla \times \mathbf{B}(\mathbf{r}, t), \\ \frac{\partial \mathbf{B}(\mathbf{r}, t)}{\partial t} &= -\nabla \times \mathbf{E}(\mathbf{r}, t). \end{aligned} \tag{3.41}$$

By differentiating these equations with respect to time, it becomes apparent that either $\mathbf{E}$ or $\mathbf{B}$ can be eliminated. For example, taking the time derivative of the second equation in (3.41):

$$\frac{\partial^2 \mathbf{B}(\mathbf{r}, t)}{\partial t^2} = -\nabla \times \underline{\frac{\partial \mathbf{E}(\mathbf{r}, t)}{\partial t}}.$$

Substituting the first equation in (3.41) into the underlined term yields:

$$\begin{aligned} \frac{\partial^2 \mathbf{B}(\mathbf{r}, t)}{\partial t^2} &= -\nabla \times \left[\frac{1}{\varepsilon_0\mu_0} \nabla \times \mathbf{B}(\mathbf{r}, t) \right], \\ \therefore \quad -\varepsilon_0\mu_0 \frac{\partial^2 \mathbf{B}(\mathbf{r}, t)}{\partial t^2} &= \nabla \times [\nabla \times \mathbf{B}(\mathbf{r}, t)]. \end{aligned}$$

Using the vector identity for double rotations (Sect. 9.3.2):

$$\nabla \times [\nabla \times \mathbf{B}(\mathbf{r}, t)] = -\nabla^2 \mathbf{B} + \nabla \cdot [\nabla \cdot \mathbf{B}(\mathbf{r}, t)],$$

and noting from Maxwell's equations that $\nabla \cdot \mathbf{B}(\mathbf{r}, t) = 0$, we obtain:

$$\frac{\partial^2 \mathbf{B}(\mathbf{r}, t)}{\partial t^2} = \frac{1}{\varepsilon_0 \mu_0} \nabla^2 \mathbf{B}(\mathbf{r}, t).$$

This equation takes the form of the wave equation (2.101). Comparing the two equations, the propagation speed v of the wave $\mathbf{B}$ is identified as:

$$\varepsilon_0 \mu_0 = v^{-2}.$$

The above derivation eliminated **E** to obtain an equation for **B**. Conversely, eliminating **B** to derive an equation for **E** yields:

$$\frac{\partial^2 \mathbf{E}(\mathbf{r}, t)}{\partial t^2} = \frac{1}{\varepsilon_0 \mu_0} \nabla^2 \mathbf{E}(\mathbf{r}, t).$$

Thus, it is also concluded that the electric field **E** is a wave with the same propagation speed. The propagation speed of electromagnetic waves is conventionally denoted as c, and hence:

$$c = 1/\sqrt{\varepsilon_0 \mu_0}.$$

The propagation of these **electromagnetic waves** was later identified with light (Sect. 5.1.1). Consequently, c represents the speed of light, and the above equation reveals that the product of the vacuum permittivity and permeability, introduced in Coulomb's law and Biot-Savart's law, respectively, is intrinsically linked to the speed of light.

When we say "electric and magnetic fields propagate as electromagnetic waves at the speed of light", a concrete image is as shown in Fig. 3.5. Recalling the wave

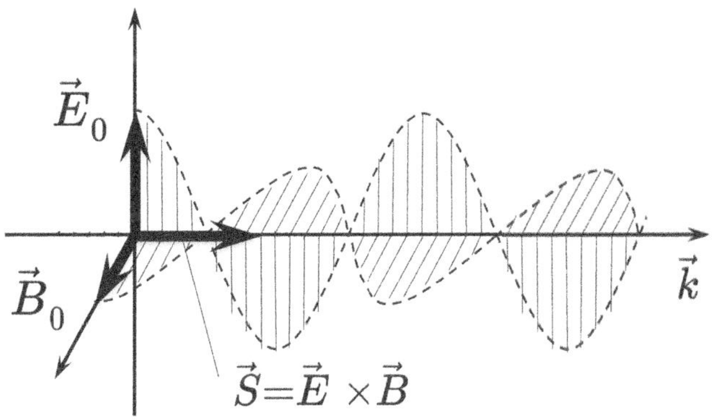

Fig. 3.5 A concrete image of electromagnetic waves can be associated with the Poynting vector $\mathbf{S} = \mathbf{E} \times \mathbf{B}$, which corresponds to light rays in geometrical optics

expression (2.105), the electric field $\mathbf{E}(\mathbf{r}, t)$ can be expressed as

$$\mathbf{E}(\mathbf{r}, t) \sim \mathbf{E}_0 \cdot e^{i(\mathbf{k} \cdot \mathbf{r} - \omega t)},$$

and similarly for the magnetic field $\mathbf{B}(\mathbf{r}, t)$.[25] Substituting this into Eq. (3.41), the time and spatial derivatives transform into

$$-i\omega \cdot \mathbf{E}_0 \sim i\mathbf{k} \times \mathbf{B}_0.$$

Here, $\mathbf{k}$ represents the propagation direction of the wave, as noted in Eq. (2.104). This cross-product relationship implies that the wave's propagation direction $\mathbf{k}$ is orthogonal to the electric field's amplitude $\mathbf{E}_0$ and the magnetic field's amplitude $\mathbf{B}_0$, as depicted in Fig. 3.5. The vector

$$\mathbf{S} = \mathbf{E} \times \mathbf{B}$$

points in the same direction as $\mathbf{k}$ and is known as the Poynting vector. This vector corresponds to the light ray in geometrical optics (Sect. 5.1.1).

References

1. Mansfield MM, O'Sullivan C (2020) Understanding physics, 3rd edn. Wiley, Hoboken. ISBN: 978-1119519508
2. Fleisch D (2011) A student's guide to vectors and tensors. Cambridge University Press, Cambridge. ISBN: 978-0521171908
3. Feynman R, Morinigo F, Wagner W (2002) Feynman lectures on gravitation. CRC Press, Boca Raton. ISBN: 978-0813340388
4. Panofsky WKH, Phillips M (2005). Classical electricity and magnetism, 2nd edn. Dover, New York. ISBN: 978-0486439242
5. Roche JJ (1998) The mathematics of measurement. Springer, Berlin. ISBN: 978-0387915814
6. Stevens CF (1995) The six core theories of modern physics. Bradford Books, Cambridge. ISBN: 978-0262193597
7. Flanders H (1989) Differential forms with applications to the physical sciences. Dover, New York. ISBN: 978-0486661698
8. Landau LD (1984) Electrodynamics of continuous media. Butterworth-Heinemann, Oxford. ISBN: 978-0750626347
9. Born M, Wolf E (2019) Principles of optics (60th Anniversary Edition). Cambridge University Press, Cambridge. ISBN: 978-1108477437
10. Jackson JD (1998) Classical electrodynamics. Wiley, New York. ISBN: 978-0471309321
11. Muta T (2016) Denji Rikigaku (in Japanese). Iwanami, Tokyo. ISBN: 978-4007304613
12. Kotani M (1954) Denjikigaku (in Japanese). Iwanami, Tokyo

[25] This considers a general term of the Fourier expansion (Sect. 2.4.5) describing the general time-space dependence of $\mathbf{E}(\mathbf{r}, t)$, as in Eq. (2.105).

Chapter 4
Key Points of Mechanics

In this chapter, we explain the key points in the framework of analytical mechanics. This section can be regarded as the "core of the book" and serves as the "trailhead" for beginners venturing into quantum physics. Some readers find it easier to understand quantum mechanics using an axiomatic or formal scientific style, namely, understanding it "from the rules". However, there are others who find it more intuitive to trace "how the ideas were originally developed" (Sect. 1.4.1). For the latter group, introducing wave mechanics through the optical-mechanical analogy from the Hamilton-Jacobi formulation of analytical mechanics provides the "most seamless trail" for the ascent. The content of this chapter provides the minimum path necessary to reach the Hamilton-Jacobi formulation. Furthermore, the material covered up to Chap. 2 can be considered the "pre-hike preparations" required before tackling this chapter.

4.1 Conservation Laws

Quantities such as energy and momentum play crucial roles in the subsequent discussions. For students without prior exposure to elementary physics, it is necessary to introduce these quantities through various conservation laws.

In high school physics, **Newton's equation of motion**[1] is given by

$$m\frac{d^2\mathbf{r}}{dt^2} = \mathbf{F}, \tag{4.1}$$

[1] Sir Isaac Newton (1643.1–1727.3, England).

R. Maezono, *Essentials for Deeper Understanding of Quantum Computing*,
https://doi.org/10.1007/978-981-96-5646-2_4

where m is the mass of the particle in question and $\mathbf{r}$ is its position [1].[2] From this equation, various conservation laws can be derived as follows:

- Transforming the equation $\rightarrow$ Conservation of momentum,
- Taking the cross product with $\mathbf{r}$ $\rightarrow$ Conservation of angular momentum,
- Taking the dot product with $(d\mathbf{r}/dt)$ $\rightarrow$ Conservation of energy.

The following sections will explain how these conservation laws are derived.

4.1.1 Conservation of Momentum

Transforming Eq. (4.1) yields

$$\frac{d}{dt}\left(m\frac{d\mathbf{r}}{dt}\right) = \mathbf{F}.$$

We define

$$m\frac{d\mathbf{r}}{dt} =: \mathbf{p},$$

calling $\mathbf{p}$ the **momentum**. Thus, the equation of motion describes the "time variation of momentum" as

$$\frac{d\mathbf{p}}{dt} = \mathbf{F}. \tag{4.2}$$

If no external force acts,

$$\frac{d\mathbf{p}}{dt} = 0, \quad \therefore \quad \mathbf{p} = const.$$

This leads to the **conservation of momentum**.

4.1.2 Conservation of Angular Momentum

Taking the cross product of both sides of Eq. (4.1) with the position vector $\mathbf{r}$, we get

$$\underline{\mathbf{r}\times} m\frac{d^2\mathbf{r}}{dt^2} = \underline{\mathbf{r}\times} \mathbf{F}. \tag{4.3}$$

[2] This notation is typically used in high school physics.

Now, the left-hand side becomes

$$\begin{aligned}(\text{LHS}) &= \mathbf{r} \times m\frac{d^2\mathbf{r}}{dt^2} \\ &= \left[\frac{d}{dt}\left(\mathbf{r} \times m\frac{d\mathbf{r}}{dt}\right) - \underline{\frac{d\mathbf{r}}{dt} \times m\frac{d\mathbf{r}}{dt}}\right] \\ &= \frac{d}{dt}\left(\mathbf{r} \times m\frac{d\mathbf{r}}{dt}\right) \\ &= \frac{d}{dt}(\mathbf{r} \times \mathbf{p}) \\ &= \frac{d\mathbf{l}}{dt}.\end{aligned}$$

Here, the underlined term vanishes because it is the cross product of a vector with itself, as noted in Eq. (2.26). We define $\mathbf{l} := \mathbf{r} \times \mathbf{p}$ as the **angular momentum**. Substituting this into Eq. (4.3), we have

$$\frac{d\mathbf{l}}{dt} = \mathbf{r} \times \mathbf{F}. \tag{4.4}$$

This equation represents the equation of motion for angular momentum, also known as the torque equation. The right-hand side, which drives the time evolution of angular momentum, is called the **torque** or the "moment of force".[3]

If the force direction coincides with the "radial direction connecting the origin to the observation point",

$$\mathbf{F} = F \cdot \mathbf{e_r},$$

this force field is called a **central force field.**[4] In this case, the right-hand side of Eq. (4.4) becomes

$$\mathbf{r} \times \mathbf{F} = F\,(\mathbf{r} \times \mathbf{e_r}) = 0,$$

since the cross product of a vector with itself is zero (see Eq. (2.26)). Therefore,

$$\frac{d\mathbf{l}}{dt} = 0, \quad \therefore \quad \mathbf{l} = const.$$

This is the **conservation of angular momentum.**

[3] The continuous version of this equation forms the foundation of rigid body motion [1].

[4] An example of this is the motion of the Earth under the Sun's gravitational force, where the Sun is at the origin and the Earth's position is **r**.

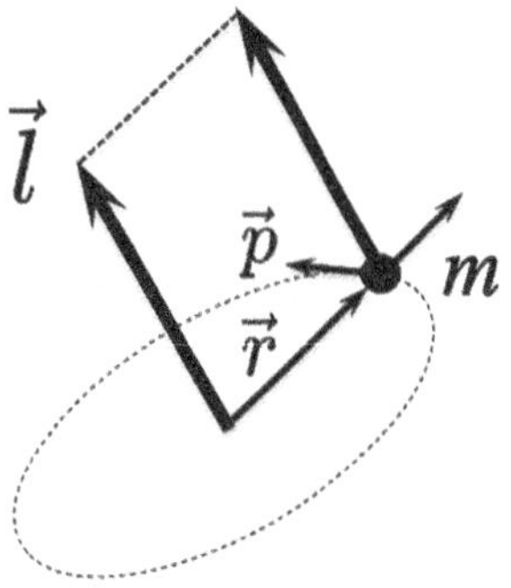

Fig. 4.1 When a mass m undergoes circular motion, an angular momentum vector **l** arises, directed along the normal to the plane of rotation. When this angular momentum is conserved, it tends to preserve its direction as a vector, making it difficult to change the orientation of the plane of rotation

Students Often Do Not Fully Understand Conservation of Angular Momentum

Note that angular momentum $\mathbf{l} = \mathbf{r} \times \mathbf{p}$, as a vector quantity, **also preserves its direction**. The angular momentum of a spinning wheel points along the wheel's axis because the "moving mass points" travel along the wheel's circumference (see Fig. 4.1). A bicycle remains upright because the direction of **l** is preserved, resisting changes to its orientation. When flying in an airplane, you may not notice a turn if you don't look outside. However, the pilot must understand the aircraft's tilt relative to the horizon. If you spin a gyroscope vertically while the aircraft is parked on the ground, the gyroscope will remember the vertical direction due to the conservation of **l**. Even after takeoff and during turns, the gyroscope continues to point vertically, helping determine the aircraft's tilt. This is the principle behind the measuring device known as a **gyroscope**.

4.1.3 Conservation of Energy

Next, taking the dot product of both sides of the equation of motion (4.1) with the velocity vector, we get

$$\frac{d\mathbf{r}}{dt} \cdot m\frac{d^2\mathbf{r}}{dt^2} = \frac{d\mathbf{r}}{dt} \cdot \mathbf{F}. \tag{4.5}$$

For the left-hand side, we have

$$\frac{d\mathbf{r}}{dt} \cdot m\frac{d^2\mathbf{r}}{dt^2} = \frac{1}{2}\frac{d}{dt}\left(\frac{d\mathbf{r}}{dt} \cdot m\frac{d\mathbf{r}}{dt}\right) = \frac{d}{dt}\left(\frac{m}{2}\left(\frac{d\mathbf{r}}{dt}\right)^2\right) =: \frac{d}{dt}T,$$

where the **kinetic energy** T is defined as

$$T = \frac{m}{2}\left(\frac{d\mathbf{r}}{dt}\right)^2 .$$

Substituting this into Eq. (4.5), we obtain

$$\frac{d}{dt}T = \frac{d\mathbf{r}}{dt} \cdot \mathbf{F}. \tag{4.6}$$

When the force is a gradient force,

$$\mathbf{F} = -\nabla U(\mathbf{r}),$$

the right-hand side of Eq. (4.6) becomes

$$\frac{d\mathbf{r}}{dt} \cdot \mathbf{F} = -\frac{d\mathbf{r}}{dt} \cdot \nabla U(\mathbf{r}) = -\frac{dr^l}{dt} \cdot \frac{\partial}{\partial r^l} U(\mathbf{r}) = -\frac{dU(\mathbf{r})}{dt},$$

by the chain rule.[5] Thus, we have

$$\frac{d}{dt}T = -\frac{dU(\mathbf{r})}{dt}, \quad \therefore \quad \frac{d}{dt}(T+U) = 0, \quad \therefore \quad E = T + U = const.$$

This leads to the **law of conservation of energy**, which states that the sum of the kinetic energy T and the potential energy U (the total energy E) remains constant.

4.2 Outline of Variational Methods

Variational methods are crucial concepts that frequently appear in the discussions that follow. In this section, we introduce the essential aspects of variational methods using the well-known textbook example of the *brachistochrone problem*.[6] We will derive only the minimum material necessary for understanding variational methods later.

4.2.1 *Stationary Value Problem of a Functional*

Consider the situation shown in Fig. 4.2, where a curve $f(x)$ is set from point P to point Q, and a ball rolls along $f(x)$ from P to Q. When the shape of the descending

[5] See Eq. (2.97).

[6] This problem was posed by Johann Bernoulli in 1696. It was shown that the brachistochrone curve is a cycloid.

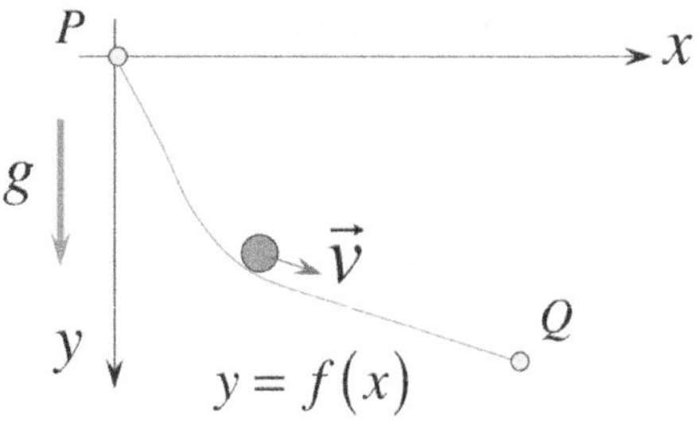

Fig. 4.2 The Brachistochrone problem. A curve $f(x)$ is set from point P to point Q. A ball rolls along this curve from P to Q

surface $f(x)$ changes, the time T required to reach the destination also changes accordingly. In this sense, we write $T = T[f]$ and say, "T is a **functional** of $f(x)$".[7] The difference between functions and functionals will be discussed in detail later.

How should $f(x)$ be chosen so that $T[f]$ is minimized, i.e., so that the ball reaches Q from P in the shortest possible time? This is the "brachistochrone problem". The approach to solving this problem will be explained later. For now, let's recall high school knowledge and derive the specific expression for $T = T[f]$. First, if we denote the distance along the curve $f(x)$ as s, the expression for the ball's speed $v(s)$ gives:

$$v(s) = \frac{ds}{dt} \quad \rightarrow \quad dt = \frac{ds}{v(s)}.$$

Hence, the time required can be expressed as an integral over the distance s:

$$T = \int dt = \int \frac{ds}{v(s)}. \tag{4.7}$$

If we take the y-coordinate pointing downward from point P, as shown in Fig. 4.2, the conservation of energy, "(kinetic energy) + (potential energy) = constant", yields:

[7] The functional T of the function $f(x)$ is written as $T[f]$, and it is standard notation not to write the x-dependence of $f(x)$. However, when this notation is introduced to beginners who are unfamiliar with functionals, one often notices that they confuse the difference between the function $T(f)$ and the functional $T[f]$. For this reason, it is sometimes intentionally written as $T[f(x)]$ to remind them that it is a functional determined by the function $f(x)$. Why, then, is $T[f]$ considered the standard notation? There is a good reason for this that also takes beginners into consideration: When performing functional differentiation with respect to $f(x)$, some beginners mistakenly think that terms such as $f(x')$ or $f(x'')$ inside the functional are not subject to the differentiation operation because the variables x' and x'' are "different from x". In functional differentiation, the notation of dummy variables such as x' is irrelevant, and all occurrences of f are subject to the operation. To avoid such confusion, the standard notation $T[f]$ is used, without writing the dummy variable explicitly.

$$\frac{1}{2}mv^2(s) = m \cdot g \cdot y \quad \rightarrow \quad v(s) \propto \sqrt{y} = \sqrt{f(x)}. \tag{4.8}$$

Thus, $v(s)$ can be evaluated. From high school knowledge, the line element ds along the curve $f(x)$ is given by:[8]

$$ds^2 = dx^2 + dy^2$$

$$\rightarrow \quad ds = \sqrt{dx^2 + dy^2} = dx\sqrt{1 + \left(\frac{dy}{dx}\right)^2} = \sqrt{1 + f'(x)^2} \cdot dx. \tag{4.9}$$

Substituting Eqs. (4.8) and (4.9) into Eq. (4.7), we get:

$$T = \int_0^{Q_x} \frac{\sqrt{1 + f'(x)^2}}{\sqrt{f(x)}} \cdot dx = T[f]. \tag{4.10}$$

Thus, we have explicitly expressed the dependence of $T[f]$ on $f(x)$.

Please note that the brackets $[f]$ used to represent a functional, unlike the parentheses (f) used to represent a function, indicate dependence not only on the values $f(x)$ themselves but also on quantities such as the derivative $f'(x)$. A functional expresses a relationship that is determined by $f(x)$ but goes beyond just the function's values themselves.

4.2.2 *How to Achieve Optimizing Functionals*

Our problem of finding the *brachistochrone* was formulated as: "Find a function $f(x)$ that minimizes the functional $T[f(x)]$ given by Eq. (4.10)". This is known as "finding the stationary value of the functional $T[f(x)]$", or the **functional stationary value problem**. Various formulas for solving such problems can be found in textbooks on physical and applied mathematics [2]. However, the logic behind *how* these formulas are derived will appear repeatedly, so it is important to explain this carefully.

The problem of 'minimizing something' resembles the idea from high school mathematics: "To find the maxima or minima of a function $g(x)$, solve $g'(x) = 0$". Reconsidering how this conclusion is derived provides a hint for solving the problem of "finding $f(x)$ that minimizes $T[f(x)]$". When finding the extrema of a function, you might intuitively understand that "the slope is zero at extrema" by drawing the graph of $g(x)$. However, this geometric intuition is not general enough to handle the problem of finding stationary values of a functional. Therefore, we will adopt a different approach.

[8] Refer to the column in Sect. 2.4.7.

Suppose $g(x)$ has a minimum at $x = x^*$. In the vicinity of $x = (x^* + \delta x)$, the behavior of $g(x)$ can be expanded as

$$g(x) = g(x^*) + g'(x^*) \cdot \delta x + O(\delta x^2). \tag{4.11}$$

(See Sect. 2.1.2). Now, we consider how to express "having a minimum" mathematically. This condition can be restated as: "whether we shift right ($\delta x > 0$) or left ($\delta x < 0$), $g(x)$ always increases". Since $g'(x^*)$ is a constant, the term $g'(x^*) \cdot \delta x$ should change its sign depending on the sign of δx. This contradicts the requirement that $g(x)$ must always increase. Therefore, the term $g'(x^*) \cdot \delta x$ must be zero. Thus, we conclude that x^* is a minimum point if it satisfies the condition $g'(x^*) = 0$.

Applying this idea allows us to approach the problem of "finding $f(x)$ that minimizes the functional $T[f]$". Let the optimal solution be $f^*(x)$, and let the deviation from it be $\delta f(x)$. Then we can express

$$T[f^* + \delta f] = T[f^*] + \delta T.$$

We now evaluate the remainder term δT. The condition that "$f^*(x)$ gives a minimum of $T[f]$" means that δT must always increase, regardless of the sign of $\delta f(x)$. Therefore, expanding $T[f^* + \delta f]$ gives

$$T[f^* + \delta f] = T[f^*] + C^{(1)} \cdot \delta f + O(\delta f^2). \tag{4.12}$$

For $f^*(x)$ to be a minimum, the coefficient of the first-order term $C^{(1)}$ must be zero. This condition mathematically expresses the notion of a minimum. Thus, by solving the equation $C^{(1)} = 0$, we can determine $f^*(x)$. This is the fundamental concept behind solving optimization problems using the variational method.

Next, we proceed to discuss how this concept applies to specific individual cases. When T depends only on f and f', as in Eq. (4.10), it can be expressed as

$$T[f] = \int_a^b dx \cdot \Phi(f(x), f'(x)). \tag{4.13}$$

In this case, the coefficient $C^{(1)}$ in Eq. (4.12) is given by

$$C^{(1)} = \frac{\partial \Phi}{\partial f} - \frac{d}{dx}\left(\frac{\partial \Phi}{\partial f'}\right),$$

as derived later in the next subsection. Setting this equal to zero leads to the equation

$$\frac{\partial \Phi}{\partial f} - \frac{d}{dx}\left(\frac{\partial \Phi}{\partial f'}\right) = 0. \tag{4.14}$$

This equation provides the minimizer $f^*(x)$ and is called the **Euler-Lagrange equation.**[9] For the example in Eq. (4.10), substituting

$$\Phi\left(f(x), f'(x)\right) = \frac{\sqrt{1+f'(x)^2}}{\sqrt{f(x)}},$$

yields a specific differential equation that $f(x)$ must satisfy. Solving this equation determines the brachistochrone curve $f^*(x)$. This is the logical flow for finding the solution.

In the extremum problems for functions studied in high school, the condition $g'(x^*) = 0$ provides an *equation for the value* x^*. In contrast, for the stationary value problems of functionals, the stationary condition gives a *differential equation that the function* $f^*(x)$ must satisfy. Pay attention to this contrast between the two cases.

4.2.3 Euler-Lagrange Equation

Now, let us derive Eq. (4.14). Consider a functional of the form in Eq. (4.13):

$$T\left[f^* + \delta f\right] = \int_a^b dx \cdot \Phi\left(f^* + \delta f, f'^* + \delta f'\right).$$

Using the linear approximation (2.6) from Sect. 2.1.2, the integrand expands to

$$\Phi\left(f^* + \delta f, f'^* + \delta f'\right) = \Phi\left(f^*, f'^*\right) + \frac{\partial \Phi}{\partial f}\delta f + \frac{\partial \Phi}{\partial f'}\delta f' + O\left(\delta^2 f\right).$$

Substituting this into the functional, we get

$$\begin{aligned} T\left[f^* + \delta f\right] &= \int_a^b dx \left[\Phi\left(f^*, f'^*\right) + \frac{\partial \Phi}{\partial f}\delta f + \frac{\partial \Phi}{\partial f'}\delta f'\right] + O\left(\delta^2 f\right), \\ &= T\left[f^*\right] + \int_a^b dx \left[\frac{\partial \Phi}{\partial f}\delta f + \frac{\partial \Phi}{\partial f'}\delta f'\right] + O\left(\delta^2 f\right). \end{aligned}$$

The first-order variation $\delta T^{(1)}$ is therefore

$$\delta T^{(1)} = \int_a^b dx \left[\frac{\partial \Phi}{\partial f}\delta f + \frac{\partial \Phi}{\partial f'}\delta f'\right] = \int_a^b dx \frac{\partial \Phi}{\partial f}\delta f + \delta T_{f'}. \tag{4.15}$$

---

[9] Joseph-Louis Lagrange (1736.1–1813.4, France).

The second term is

$$\delta T_{f'} = \int_a^b dx \left[\frac{\partial \Phi}{\partial f'} \delta f' \right].$$

Now, since

$$\delta f' = f' - f'^* = \frac{df}{dx} - \frac{df^*}{dx} = \frac{d}{dx}(f - f^*) = \frac{d\delta f}{dx},$$

we have

$$\delta T_{f'} = \int_a^b dx \left[\frac{\partial \Phi}{\partial f'} \frac{d\delta f}{dx} \right].$$

Integrating by parts,

$$\delta T_{f'} = \left[\frac{\partial \Phi}{\partial f'} \delta f \right]_a^b - \int_a^b dx \frac{d}{dx} \left(\frac{\partial \Phi}{\partial f'} \right) \delta f.$$

If we set $\delta f(a) = \delta f(b) = 0$, the boundary term vanishes, giving

$$\delta T_{f'} = - \int_a^b dx \frac{d}{dx} \left(\frac{\partial \Phi}{\partial f'} \right) \delta f.$$

Substituting back into Eq. (4.15),

$$\begin{aligned} \delta T^{(1)} &= \int_a^b dx \frac{\partial \Phi}{\partial f} \delta f - \int_a^b dx \frac{d}{dx} \left(\frac{\partial \Phi}{\partial f'} \right) \delta f, \\ &= \int_a^b dx \left[\frac{\partial \Phi}{\partial f} - \frac{d}{dx} \left(\frac{\partial \Phi}{\partial f'} \right) \right] \delta f = C^{(1)} \cdot \delta f. \end{aligned}$$

To satisfy the stationary condition $\delta T^{(1)} = 0$ for any δf, the integrand must satisfy

$$\frac{\partial \Phi}{\partial f} - \frac{d}{dx} \left(\frac{\partial \Phi}{\partial f'} \right) = 0.$$

This is the **Euler-Lagrange equation** derived in the previous section (Eq. (4.14)).

4.3 Lagrangian Mechanics and the Action Integral

4.3.1 *Equations of Motion in Generalized Coordinates*

Newton's equations of motion are formulated in the usual Cartesian coordinate system (x, y, z). However, most practical problems involve **constraints** , such as "a particle attached to a string" or "a particle rolling on a curved surface". For example, in the case of a pendulum suspended by a string of length r, the position of the mass can be expressed as $x = r\sin\theta$ and $z = r\cos\theta$. To describe the motion, we need to convert the equations of motion from (x, z) to the coordinates (r, θ) and reframe them in terms of the dynamics with respect to the angle θ. Thus, it becomes necessary to express the laws of motion using coordinate systems appropriate to the problem, known as **generalized coordinates**. The desire to systematize this conversion into a general procedure is the starting motivation for the discussion in this section.

Let us denote the generalized coordinates as $\{q_j\}$. For example, in the case of spherical coordinates, (r, θ, ϕ) corresponds to (q_1, q_2, q_3), and we have:

$$\begin{aligned} x_1 &= x = r\sin\theta\cos\phi = q_1 \sin q_2 \cos q_3, \\ x_2 &= y = r\sin\theta\sin\phi = q_1 \sin q_2 \sin q_3, \\ x_3 &= z = r\cos\theta = q_1 \cos q_2. \end{aligned}$$

In general, such relationships can be written as:

$$x_i = x_i\left(q_1, \cdots, q_M\right).$$

Writing Newton's equation of motion as in Eq. (4.2) and considering a gradient force $\mathbf{F}(\mathbf{r}) = -\nabla U(\mathbf{r})$, the equation in component form becomes:

$$\frac{dp_j}{dt} = -\frac{\partial U}{\partial x_j}. \tag{4.16}$$

Consider that (x_j, p_j) can be expressed in terms of generalized coordinates as:

$$x_i = x_i\left(q_1, \cdots, q_M\right), \quad p_i = p_i\left(q_1, \cdots, q_M\right). \tag{4.17}$$

Newton's equation of motion describes a "drive equation for momentum", expressing how the time derivative of momentum is determined. Therefore, our goal is to proceed along the following lines:

1. Introduce a generalized definition of momentum,
2. Describe the laws of motion in terms of how the time derivative of this *generalized momentum* is determined.

First, regarding the **generalization of momentum**, in the case of a Cartesian coordinate system, the kinetic energy is given by:

$$T = \sum_j \frac{m}{2} \dot{x}_j^2.$$

Taking the derivative with respect to $\dot{x}_j$, we have:

$$\frac{\partial T}{\partial \dot{x}_j} = m\dot{x}_j = p_j.$$

Thus, momentum can be expressed as:

$$p_j = \frac{\partial T}{\partial \dot{x}_j}. \tag{4.18}$$

Taking this relation as the generalized definition, we make for generalized coordinates $\{q_j\}$ to define **generalized momentum** $\tilde{p}_j$ as:

$$\tilde{p}_j = \frac{\partial T}{\partial \dot{q}_j}.$$

With this definition, in the Cartesian coordinate system, the generalized momentum reduces to the usual momentum.

Following the approach outlined in step (2), we now evaluate

$$\frac{d\tilde{p}_j}{dt} = \cdots \tag{4.19}$$

to derive the equation describing the time evolution of generalized momentum and, consequently, express the *equation of motion* in terms of generalized coordinates. Through this evaluation, we will obtain a formulation of the laws of motion that generalizes the drive equation for momentum. Using the fact that the kinetic energy T in a Cartesian coordinate system depends only on $\{\dot{x}_i\}$, we have:[10]

$$\tilde{p}_j = \frac{\partial T}{\partial \dot{q}_j} = \frac{\partial T(\{\dot{x}_i\})}{\partial \dot{q}_j} = \sum_l \frac{\partial T}{\partial \dot{x}_l} \cdot \frac{\partial \dot{x}_l}{\partial \dot{q}_j} = \sum_l p_l \cdot \frac{\partial \dot{x}_l}{\partial \dot{q}_j}. \tag{4.20}$$

To evaluate $\partial \dot{x}_l / \partial \dot{q}_j$ to be substituted, we differentiate the first equation in Eq. (4.17) with respect to time:

$$\dot{x}_i = \frac{dx_i(\{q_l\})}{dt} = \sum_l \frac{\partial x_i}{\partial q_l} \cdot \frac{dq_l}{dt} = \sum_l \frac{\partial x_i}{\partial q_l} \cdot \dot{q}_l.$$

[10] This applies the chain rule described in Sect. 2.4.7 and the momentum expression in Eq. (4.18).

Differentiating this expression with respect to $\dot{q}_j$ yields:

$$\frac{\partial \dot{x}_l}{\partial \dot{q}_j} = \frac{\partial x_l}{\partial q_j}.$$

Thus, substituting back into Eq. (4.20), we obtain:

$$\tilde{p}_j = \sum_l p_l \cdot \frac{\partial x_l}{\partial q_j}.$$

Thus, the drive equation in the form of Eq. (4.19) is evaluated as:

$$\frac{d\tilde{p}_j}{dt} = \sum_l \frac{d}{dt}\left(p_l \cdot \frac{\partial x_l}{\partial q_j}\right) = \sum_l \left(\dot{p}_l \cdot \frac{\partial x_l}{\partial q_j} + p_l \cdot \frac{\partial \dot{x}_l}{\partial q_j}\right).$$

Substituting Eqs. (4.18) and (4.16):

$$p_j = \frac{\partial T}{\partial \dot{x}_j}, \quad \frac{dp_j}{dt} = -\frac{\partial U}{\partial x_j},$$

and applying the chain rule for index contraction (Sect. 2.4.7), we have:

$$\begin{aligned}\frac{d\tilde{p}_j}{dt} &= \sum_l \left(\left(-\frac{\partial U}{\partial x_l}\right) \cdot \frac{\partial x_l}{\partial q_j} + \frac{\partial T}{\partial \dot{x}_l} \cdot \frac{\partial \dot{x}_l}{\partial q_j}\right), \\ &= \frac{\partial T}{\partial q_j} - \frac{\partial U}{\partial q_j}, \\ &= \frac{\partial}{\partial q_j}(T - U).\end{aligned}$$

Defining the quantity **Lagrangian** L as:

$$L := T - U, \tag{4.21}$$

we obtain the desired *drive equation for the generalized momentum*:

$$\frac{d\tilde{p}_j}{dt} = \frac{\partial L}{\partial q_j}.$$

This expresses the *time evolution of the generalized momentum.*

In summary, the equation of motion in Cartesian coordinates:

$$\frac{dp_j}{dt} = -\frac{\partial U}{\partial x_j}$$

is rewritten in generalized coordinates as:

$$\frac{d\tilde{p}_j}{dt} = \frac{\partial L}{\partial q_j}. \tag{4.22}$$

This represents the equation of motion in terms of the generalized coordinates and the Lagrangian.

4.3.2 *Lagrange's Equation of Motion and the Principle of Least Action*

The potential U, by its definition in Sect. 3.1.6, depends only on the coordinates. Therefore, it can be expressed as $U\left(\{q_j\}\right)$. On the other hand, the kinetic energy T in Cartesian coordinates depends solely on the time derivatives of the coordinates. However, in generalized coordinates, the kinetic energy may also depend on the coordinates themselves. Therefore, the Lagrangian's arguments are:

$$L = T\left(\{\dot{q}_j\}, \{q_j\}\right) - U\left(\{q_j\}\right) = L\left(\{q_j, \dot{q}_j\}\right). \tag{4.23}$$

This confirms that the Lagrangian L is a function of both the generalized coordinates $\{q_j\}$ and their time derivatives $\{\dot{q}_j\}$. Therefore, the generalized momentum can be expressed as:

$$\tilde{p}_j = \frac{\partial T}{\partial \dot{q}_j} = \frac{\partial L}{\partial \dot{q}_j}.$$

Substituting this into the equation of motion (4.22), we get:

$$\frac{d}{dt}\left(\frac{\partial L}{\partial \dot{q}_j}\right) = \frac{\partial L}{\partial q_j}, \quad \therefore \quad \frac{\partial L}{\partial q_j} - \frac{d}{dt}\left(\frac{\partial L}{\partial \dot{q}_j}\right) = 0.$$

This equation, expressed purely in terms of the Lagrangian L and the generalized coordinates $\{q_j\}$, is called the **Lagrange's equation of motion**.

Covariance of the Theory
In the Lagrangian formulation of the equations of motion, no specific assumptions about the generalized coordinates in Eq. (4.17) have been made. Therefore, the equation of motion (4.22) holds for any choice of generalized coordinates. This contrasts with Newton's equations, which are formulated specifically for Cartesian coordinate systems. If the equation of motion (4.22)

(continued)

holds in one coordinate system, it remains valid even if the coordinate system is transformed. The form of the governing equation does not change under such transformations. This property is referred to as the **covariance** of the governing equation with respect to coordinate transformations. Lagrangian mechanics offers the advantage of expressing the equations of motion in a way that is independent of the choice of coordinates. Additionally, since the governing equations are formulated using scalar quantities, errors related to the choice of direction are less likely to occur. These advantages make the Lagrangian approach highly applicable in fields such as mechanical dynamics and industrial engineering [3].

Recalling the content of Sect. 4.2, minimizing the functional in Eq. (4.13):

$$T[f] = \int_a^b dx \cdot \Phi\left(f(x), f'(x)\right)$$

leads to the equation:

$$\frac{\partial \Phi}{\partial f} - \frac{d}{dx}\left(\frac{\partial \Phi}{\partial f'}\right) = 0. \tag{4.24}$$

Now, the Lagrange equation of motion is:

$$\frac{\partial L}{\partial q_j} - \frac{d}{dt}\left(\frac{\partial L}{\partial \dot{q}_j}\right) = 0. \tag{4.25}$$

Comparing Eq. (4.25) with Eq. (4.24), we recognize that the Lagrange equation of motion describes the principle of minimizing the functional integral:

$$S = \int_{t_0}^{t_1} dt \cdot L\left(\{q_j(t), \dot{q}_j(t)\}\right). \tag{4.26}$$

This integral is called the **action integral**.

The laws of motion, initially discovered in the language of Cartesian coordinates through Newtonian mechanics, have been reinterpreted through mathematical reasoning as: "Nature governs phenomena in such a way as to minimize the action integral". This is known as the **principle of least action**. This formulation mirrors **Fermat's principle** in geometrical optics, which states that light follows the path of least time. Thus, an analogy between optics and mechanics is established, forming one of the breakthroughs that eventually led to quantum mechanics (Sect. 5.1).

Geometrical Optics
In high school, you learn about the propagation laws of light, such as straight-line motion, the law of reflection, and the law of refraction. These principles form the basis of what is known as *geometrical optics*. The empirical laws of geometrical optics can be derived from the following *line integral of the refractive index* along the propagation path (Sect. 2.5.1):

$$I = \int_P^Q ds \cdot n\left(\mathbf{r}(s)\right).$$

According to *Fermat's principle*, the propagation path is determined such that this integral is minimized. From this principle, various propagation laws are deduced in geometrical optics [4].

4.4 Hamiltonian Formulation of Mechanics

4.4.1 Legendre Transformation

A function $y = y(x)$ can be interpreted as defining a "restricted state" (a subspace) in the (x, y) plane. This restriction is given in the form: "for each x, a corresponding $y(x)$ is specified", essentially describing the state as a set of points. If we restrict our attention to convex functions, we can, as shown in Fig. 4.3, reconstruct the same subspace $y(x)$ by describing it as the *envelope of tangent lines*.

Let's explore how the 'representation of a convex function $y(x)$ via its envelope of tangents' can be described quantitatively. This can be achieved by providing a set $\psi(p)$ that associates each slope p of the tangent with its intercept ψ. Specifically,

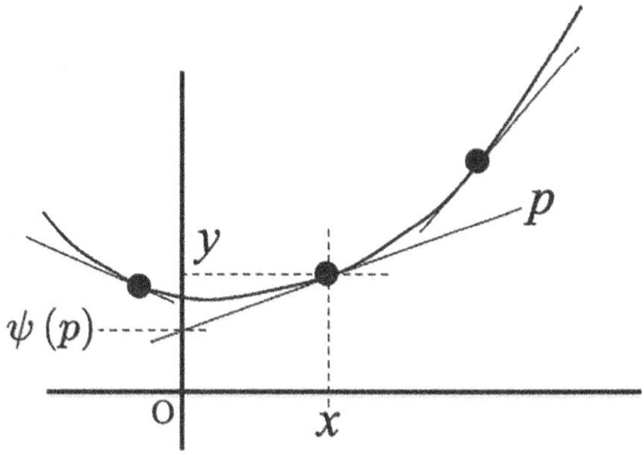

Fig. 4.3 A convex function can be depicted as the envelope of tangents by specifying the set $\psi(p)$, which represents the intercepts of tangents with slope p

we have:

$$y(x) \Leftrightarrow \psi(p), \quad p = \frac{dy}{dx}.$$

This establishes a one-to-one correspondence that describes the same subspace (the restricted "state"). In physics, it is often more convenient to define a state by the rate of change p rather than by the point x (see the column in this subsection). The former approach uses $y(x)$, while the latter uses $\psi(p)$. The relationship connecting these two representations is provided by the **Legendre transformation**. Next, we will investigate the specific relationship between $y(x)$ and $\psi(p)$.

Thermodynamics

In thermodynamics, the energy of a system is typically expressed as a function of volume, $U(V)$. The pressure of the system is then given by the derivative:

$$p = -\frac{\partial U(V)}{\partial V}.$$

In practical applications, such as in chemistry, it is sometimes more convenient to describe the state of a system in terms of the pressure p rather than the volume V. This motivation parallels the discussion in the main text, where we consider defining a state by the slope of the tangent p rather than by the point x. In thermodynamics, the approach of defining a state by the condition "at pressure p" rather than "at volume V" illustrates the utility of this alternative perspective [5].

Consider a convex function $y = y(x)$ as shown in Fig. 4.3. Defining the function as $y = y(x)$ corresponds to specifying the curve in the form: "for each point x, a corresponding point $y(x)$ is given". This approach represents the curve as a "set of points", which corresponds to *point geometry*. Now, let's consider whether this relationship $y = y(x)$ can be equivalently defined by the tangent information, specifically by the set of slopes $\left\{p(x) = \frac{dy}{dx}\right\}$. Defining the curve as an "envelope of tangents" corresponds to *line geometry*. For convex functions, each slope p corresponds to a unique point on the curve, making this approach feasible. However, a tangent with a given slope p is not uniquely defined unless its y-intercept is specified. By defining the set $\psi(p)$, which represents "the intercept of the tangent with slope p", we can reconstruct the original curve $y = y(x)$ as the envelope of these tangents.

From Fig. 4.3, we have the relationship:

$$p = \frac{y(x) - \psi(p)}{x - 0}.$$

Therefore, we can write:

$$\psi(p) = y(x) - p \cdot x, \quad p = \frac{dy(x)}{dx}. \tag{4.27}$$

This establishes a complete transfer of information between $y(x)$ and $\psi(p)$. If we wish to specify the state using the derivative $p = dy/dx$ instead of x, Eq. (4.27) ensures that the system's state is transferred *without loss of information*. This transformation is known as the **Legendre transformation**.

4.4.2 Time-Shift Invariance and the Hamiltonian

In Sect. 4.3.2, the equations of motion for a system were described using the *Lagrange equation*. The Lagrangian, which serves as a generalized potential, is a fundamental quantity that governs these equations of motion. By requiring the invariance of the Lagrangian under translations or rotations, various important consequences can be derived.[11] Here, we introduce the **Hamiltonian** by examining the invariance of the Lagrangian with respect to *time shifts*, which leads to the conservation of energy.[12]

The time derivative of the Lagrangian is given by:[13]

$$\frac{dL(q, \dot{q})}{dt} = \frac{\partial L}{\partial t} + \frac{dq}{dt}\frac{\partial L}{\partial q} + \frac{d\dot{q}}{dt}\frac{\partial L}{\partial \dot{q}}.$$

If the Lagrangian does not explicitly depend on time, i.e.,

$$\frac{\partial L}{\partial t} = 0,$$

we get:

$$\frac{dL(q, \dot{q})}{dt} = \frac{dq}{dt}\underline{\frac{\partial L}{\partial q}} + \frac{d\dot{q}}{dt}\frac{\partial L}{\partial \dot{q}}.$$

From the Lagrange equation (4.25), we know:

$$\frac{d}{dt}\left(\frac{\partial L}{\partial \dot{q}}\right) = \frac{\partial L}{\partial q}.$$

[11] Typical explanations on this topic are found in Landau's textbook [6], and a series of results known as Noether's theorem [7], where invariance requirements lead to conservation laws.

[12] The same reasoning is provided in Landau's textbook [6], where this quantity is referred to as energy, not the Hamiltonian (see §6).

[13] Refer to Sect. 2.4.7.

Substituting this into the underlined term, we have:

$$\frac{dL(q,\dot{q})}{dt} = \dot{q}\underline{\frac{d}{dt}\left(\frac{\partial L}{\partial \dot{q}}\right)} + \frac{d\dot{q}}{dt}\frac{\partial L}{\partial \dot{q}} = \frac{d}{dt}\left(\dot{q}\frac{\partial L}{\partial \dot{q}}\right). \tag{4.28}$$

Since the generalized momentum is:

$$\frac{\partial L}{\partial \dot{q}} = p,$$

Equation (4.28) becomes:

$$\frac{dL(q,\dot{q})}{dt} = \frac{d}{dt}(\dot{q}\cdot p).$$

Therefore:

$$\frac{d}{dt}(L - \dot{q}\cdot p) = 0.$$

We introduce the **Hamiltonian** H as:

$$-H := L(q,\dot{q}) - \dot{q}\cdot p.$$

Rewriting this as:

$$-H = L(q,\dot{q}) - \dot{q}\cdot p, \quad p = \frac{\partial L}{\partial \dot{q}},$$

and comparing with Eq. (4.27), we see that the Hamiltonian is the **Legendre transformation** of the Lagrangian, where the variable $\dot{q}$ is replaced by the generalized momentum p:

$$\dot{q} \quad \leftrightarrow \quad p = \frac{\partial L}{\partial \dot{q}}.$$

Thus, the Hamiltonian H is a function of p and q:

$$-H(p,q) := L(q,\dot{q}) - \dot{q}\cdot p. \tag{4.29}$$

In a Cartesian coordinate system, we have:

$$L - \dot{q}\cdot p = T - U - \dot{q}\cdot p = \frac{p^2}{2m} - U - \frac{p}{m}\cdot p = -U - \frac{p^2}{2m}.$$

Therefore:

$$H(p, q) = U + \frac{p^2}{2m} = U + T = E.$$

This shows that the Hamiltonian reduces to the total energy E of the system.

In undergraduate quantum mechanics courses, where time is limited, the Hamiltonian is often introduced in a somewhat ad hoc manner, blurring the distinction between it and energy. Beginners may wonder: "If it represents energy, why use the term Hamiltonian at all"? However, as shown above, the Hamiltonian arises as a conserved quantity derived from the "symmetry of the system with respect to time translations". In a standard Cartesian coordinate system, the Hamiltonian surely coincides with the energy. However, in generalized coordinates, this is not necessarily the case. The Hamiltonian should be viewed as a *generalization of energy*, specifically as the quantity conserved due to time translation symmetry [8] On a more advanced level, imposing symmetries on the Lagrangian, such as spatial translation symmetry, rotational symmetry, and time translation symmetry, leads to various conservation laws (e.g., Noether's theorem [7]). Historically, fundamental equations were empirically discovered. Modern physics, however, takes the reverse approach: by requiring symmetries in the Lagrangian, new fundamental equations are derived. These equations predict observable phenomena, which are then verified through experiments.

4.4.3 Hamilton's Equations of Motion

In the **Lagrangian formulation** described in Sect. 4.3, the fundamental variables are $(q, \dot{q})$, and the equation governing the time evolution of the dynamical variable $q(t)$ is the Lagrange equation. In contrast, the Hamiltonian formulation is obtained through a Legendre transformation that treats $p = \dot{q}$ as an *independent variable* separate from q. In this case, the governing equations equivalent to the Lagrange equations are expressed as equations controlling the "time evolution of both p and q". We will now derive these equations.

The Hamiltonian is defined as:

$$H = \dot{q} \cdot p - L(q, \dot{q}).$$

Taking the *total differential* of H:

$$\begin{aligned} dH &= p \cdot d\dot{q} + \dot{q} \cdot dp - \left(\frac{\partial L}{\partial q} dq + \frac{\partial L}{\partial \dot{q}} d\dot{q} \right), \\ &= p \cdot d\dot{q} + \dot{q} \cdot dp - \left(\frac{\partial L}{\partial q} dq + p \cdot d\dot{q} \right), \\ &= \dot{q} \cdot dp - \frac{\partial L}{\partial q} dq. \end{aligned}$$

From the Lagrange equation and the definition of generalized momentum, we know:

$$\frac{\partial L}{\partial q} = \frac{d}{dt}\left(\frac{\partial L}{\partial \dot{q}}\right) = \dot{p}.$$

Thus, we have:

$$dH = \dot{q} \cdot dp - \dot{p} dq.$$

Since the Hamiltonian H depends on the variables (p, q), we can write:[14]

$$dH(p, q) = \frac{\partial H}{\partial q} dq + \frac{\partial H}{\partial p} dp.$$

Comparing the two expressions for dH, we obtain:

$$\dot{p} = -\frac{\partial H}{\partial q}, \quad \dot{q} = \frac{\partial H}{\partial p}. \tag{4.30}$$

These equations govern the time evolution of the *independent variables* (q, p) for a given Hamiltonian. They are known as **Hamilton's equations of motion.**[15]

4.5 Time Evolution of the Action Integral; Hamilton-Jacobi Theory

In Sect. 4.3.2, we showed that the Lagrange equation describes the minimization of the action integral:

$$S = \int_{t_0}^{t_1} dt' \cdot L\left(q(t'), \dot{q}(t')\right) = S\left[q(t_1); q(t_0)\right]. \tag{4.31}$$

In this functional minimization problem, the **endpoints are fixed**, and the trajectory $q(t)$ is determined to minimize the action S. This approach asks: "Given fixed starting and ending points, what motion occurred"? Alternatively, we can consider the perspective of how the endpoint evolves over time due to the motion. In this approach, we treat the endpoint of the action integral S as a **freely evolving endpoint**:

$$S\left[q(t); q(t_0)\right] = \int_{t_0}^{t} dt' \cdot L\left(q(t'), \dot{q}(t')\right).$$

[14] Refer to Sect. 2.4.7.

[15] William Rowan Hamilton, 1805.8–1865.9, U.K.

We now aim to formulate the equations of motion that describe the "time evolution of S with respect to t" under this framework.

Let's revisit the functional minimization discussed in Sect. 4.2 by evaluating the variation of the action integral.

$$\begin{aligned} S[q+\delta q] &= \int_{t_0}^{t} dt' \cdot L(q+\delta q, \dot{q}+\delta\dot{q}) \\ &= \int_{t_0}^{t} dt' \cdot \left\{ L(q,\dot{q}) + \frac{\partial L}{\partial q}\delta q + \frac{\partial L}{\partial \dot{q}}\delta\dot{q} \right\} \\ &= S[q] + \delta S. \end{aligned} \tag{4.32}$$

The variation δS is given by:

$$\begin{aligned} \delta S &= \int_{t_0}^{t} dt' \cdot \left\{ \frac{\partial L}{\partial q}\delta q + \frac{\partial L}{\partial \dot{q}}\delta\dot{q} \right\} \\ &= \int_{t_0}^{t} dt' \cdot \left\{ \frac{\partial L}{\partial q}\delta q \right\} + \int_{t_0}^{t} dt' \cdot \left\{ \frac{\partial L}{\partial \dot{q}}\frac{d\delta q}{dt} \right\}. \end{aligned}$$

Applying integration by parts to the second term, we get:

$$\delta S = \int_{t_0}^{t} dt' \cdot \left\{ \frac{\partial L}{\partial q}\delta q \right\} + \int_{t_0}^{t} dt' \cdot \left[\underline{\frac{d}{dt}\left(\frac{\partial L}{\partial \dot{q}}\delta q \right)} - \frac{d}{dt}\left(\frac{\partial L}{\partial \dot{q}} \right)\delta q \right].$$

The underlined term integrates to:

$$\delta S = \int_{t_0}^{t} dt' \cdot \left\{ \frac{\partial L}{\partial q} - \frac{d}{dt}\left(\frac{\partial L}{\partial \dot{q}} \right) \right\}\delta q + \underline{\left[\frac{\partial L}{\partial \dot{q}}\delta q \right]_{t_0}^{t}}. \tag{4.33}$$

In the context of functional minimization in Sect. 4.2, the **endpoints are fixed**. Thus, $\delta q(t_0) = 0$ and $\delta q(t_1) = 0$, causing the final term to vanish. As a result, the integrand in the first term must be zero. This leads to the Lagrange equation of motion, as derived at the end of Sect. 4.2.3.

In this case, we treat the endpoint as a **free endpoint**, meaning that the underlined term in Eq. (4.33) is not discarded. Then we apply the fact that the integrand in the first term satisfies:

$$\frac{\partial L}{\partial q} - \frac{d}{dt}\left(\frac{\partial L}{\partial \dot{q}} \right) = 0, \tag{4.34}$$

as implying that the motion follows the rules of classical mechanics. Instead of asking, "*What motion minimizes the functional?*", we now ask, "**Given motion that follows the Lagrange equation, how does the endpoint of S evolve over time?**" From this perspective, we have:

$$\delta S = \left[\frac{\partial L}{\partial \dot{q}}\delta q\right]_{t_0}^{t}.$$

In the limit $\delta q \to 0$, the action $S = S[q(t)]$ satisfies:

$$\frac{\partial S}{\partial q} = \frac{\partial L}{\partial \dot{q}} = p. \tag{4.35}$$

This shows that the derivative of the action S with respect to q gives the generalized momentum p.

We begin by obtaining the above results and then proceed to consider the time evolution of $S = S[q(t)]$. We aim to derive a time evolution equation for S. The time derivative of S is:

$$\frac{dS[q(t)]}{dt} = \frac{\partial S}{\partial t} + \frac{\partial S}{\partial q}\frac{dq}{dt}.$$

Using Eq. (4.35), we get:

$$\frac{dS[q(t)]}{dt} = \frac{\partial S}{\partial t} + p \cdot \dot{q}. \tag{4.36}$$

On the other hand, from the definition of the action integral (4.31),

$$\frac{dS[q(t)]}{dt} = L. \tag{4.37}$$

Substituting into Eq. (4.36), we get:

$$L = \frac{\partial S}{\partial t} + p \cdot \dot{q}.$$

Therefore:

$$\frac{\partial S[q(t)]}{\partial t} = L - p \cdot \dot{q} = -H, \quad \therefore \quad \frac{\partial S[q(t)]}{\partial t} + H = 0.$$

Recalling the definition of the Hamiltonian (4.29),

$$\frac{\partial S\left[q(t)\right]}{\partial t} + H\left(q, p\right) = 0.$$

This equation, however, is not meaningful as it stands, because it mixes the Lagrangian form of S, which depends only on $q(t)$, with the Hamiltonian form of H, which depends on $(q(t), p(t))$. Since we are describing the endpoint dynamics of S, we need to eliminate p by expressing it in terms of q. From Eq. (4.35), we have:

$$p = \frac{\partial S}{\partial q}.$$

Substituting this into the Hamiltonian equation, we obtain:

$$\frac{\partial S\left[q(t)\right]}{\partial t} + H\left(q(t), \frac{\partial S}{\partial q}\right) = 0. \tag{4.38}$$

This equation describes the time evolution of the action function. It is called the **Hamilton-Jacobi equation.**[16]

In the Hamilton-Jacobi formalism, to obtain the trajectory of a particle, the process follows this sequence: First, solve the Hamilton-Jacobi equation to find the action S. Then, from the gradient of S, the momentum $\mathbf{p}$ can be obtained, which describes the trajectory.

4.6 Summary of This Chapter

Given the length of this chapter, let's review the overall structure before moving forward (Fig. 4.4). In analytical mechanics, Newtonian mechanics is reformulated in various ways, allowing us to transition between different formalisms. This approach offers multiple perspectives for revisiting classical mechanics, which is the significance of analytical mechanics. It provides a pathway to reinterpret familiar concepts and gain deeper insights into classical dynamics. It's worth noting that the "climbing route" we followed in this book is just one of many. There are other texts and courses that present different interpretations or methods of introduction. For further remarks on these alternative approaches, see Sect. 9.4.

[16] Carl Gustav Jacob Jacobi, 1804.12–1851.2, German. The Hamilton-Jacobi equation describes the "minimization of the action integral". A discrete form of this equation appears in reinforcement learning as the Bellman equation, which handles the minimization of a loss function or the maximization of a value function [9, 10].

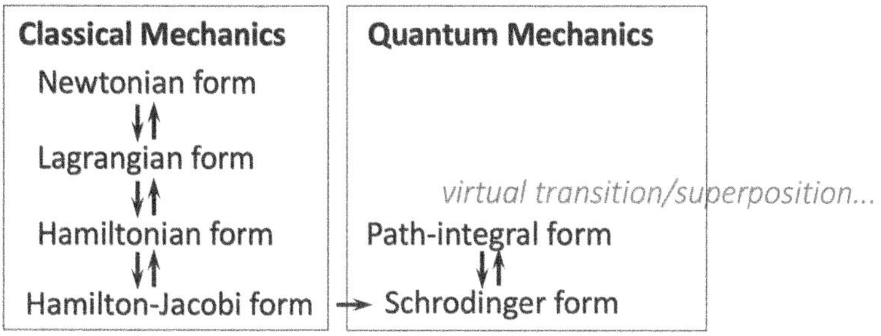

Fig. 4.4 Various formulations of mechanics. The purpose of this section is to provide an overview of these formulations based on what we have learned so far

In classical mechanics, the time evolution of a particle's position, or its **trajectory**, is described. The most elementary formulation uses the Cartesian coordinate system and the **Newtonian form**. In this approach, the time evolution of a particle's momentum $p = mv$ is described by the equation:

$$\frac{dp_j}{dt} = -\frac{\partial U\left(\{x_l\}\right)}{\partial x_j}. \tag{4.39}$$

Here, $U\left(\{q_l\}\right)$ is the potential energy.

In practical applications, the motion under consideration is often described using coordinate systems constrained by conditions in space. We need an equation of motion for such **generalized coordinates** $\left\{q_j\right\}$. However, since Newton's equation of motion (4.39) is expressed in free space and Cartesian coordinates, it is cumbersome to rewrite it each time for the desired generalized coordinates. This motivates the need for a formulation that can express the equation of motion identically for any generalized coordinates. This formulation is the **Lagrangian form**:

In the Lagrangian form, the kinetic energy is:

$$T = \sum_j \frac{m}{2}\dot{x}_j^2.$$

We introduce the generalized momentum, which is a generalized form that reduces to the usual momentum in Cartesian coordinates:

$$\tilde{p}_j = \frac{\partial T}{\partial \dot{q}_j}. \tag{4.40}$$

The equation of motion takes the form to describe the time evolution of this generalized momentum. Expressing Cartesian coordinates in terms of generalized coordinates:

$$x_i = x_i\left(q_1, \cdots, q_M\right), \quad p_i = p_i\left(q_1, \cdots, q_M\right),$$

and transforming Newton's equation of motion, we derive the equation for $d\tilde{p}_j/dt$:

$$\frac{d\tilde{p}_j}{dt} = \frac{\partial L}{\partial q_j}, \quad L = T - U. \tag{4.41}$$

Here, L is the generalized potential quantity, called the **Lagrangian**. Combining Eq. (4.41) with Eq. (4.40), we get:

$$\frac{\partial L}{\partial q_j} - \frac{d}{dt}\left(\frac{\partial L}{\partial \dot{q}_j}\right) = 0. \tag{4.42}$$

This is the equation of motion in generalized coordinates, called the Lagrangian form of the equation of motion. This equation retains the same form for any generalized coordinates, demonstrating **covariance** under coordinate transformations.

Moving away from the perspective of wanting to know the time evolution of the coordinate q, we can instead emphasize the viewpoint of the dynamics described by the second-order time derivative of q. In this perspective, the degrees of freedom reduce to the 0th-order derivative (coordinate) and the 1st-order derivative (momentum). A formalism that describes the time evolution of these two quantities is the **Hamiltonian form**. Mathematically, this transformation is realized by applying the Legendre transform to the Lagrangian, which converts it into the Hamiltonian:

$$-H\left(\mathbf{q}, \mathbf{p}\right) = L\left(\mathbf{q}, \dot{\mathbf{q}}\right) + \mathbf{p} \cdot \dot{\mathbf{q}}, \quad p_j = \frac{\partial L}{\partial \dot{q}_j}.$$

This leads to the Hamiltonian equations of motion:

$$\frac{dp_j}{dt} = -\frac{\partial H\left(\mathbf{q}, \mathbf{p}\right)}{\partial q_j}, \quad \frac{dq_j}{dt} = \frac{\partial H\left(\mathbf{q}, \mathbf{p}\right)}{\partial p_j}.$$

These equations describe the same content as Newton's equation of motion or the Lagrangian equation of motion. This formalism is known as the Hamiltonian form of the equations of motion (Fig. 4.4).

Comparing with the general framework of the variational method, the Lagrangian equation of motion (4.42) corresponds to the Euler-Lagrange equation that minimizes the integral functional:

$$S = \int_{t_0}^{t_1} dt \cdot L\left(\left\{q_j, \dot{q}_j\right\}\right).$$

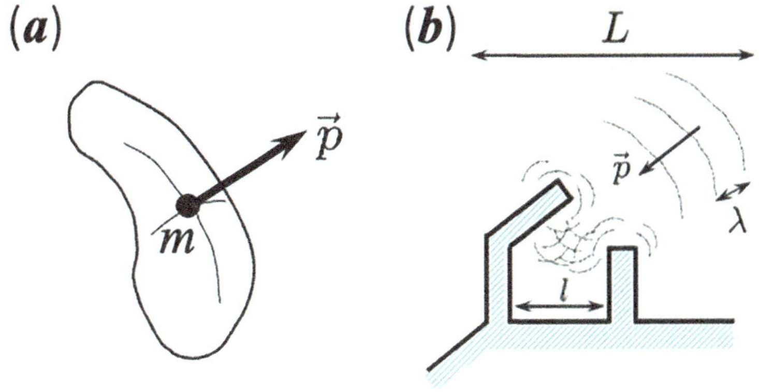

Fig. 4.5 Optical-Mechanical Analogy: The trajectory of a particle can be interpreted as the "normal to a hypothetical wavefront" [Panel (**a**)]. When a wave of wavelength λ propagates through a region, if the characteristic length of the region is significantly larger than the wavelength ($\lambda \ll L$), the wavefront normal provides a well-defined description [Panel (**b**)]. On the other hand, when the region's size is on the order of the wavelength ($L \sim \lambda$), the concept of a "wavefront normal" becomes an inadequate description

This S is called the action integral. When the time evolution of the endpoint of S:

$$S(t) = \int_{t_0}^{t} dt \cdot L\left(\left\{q_j, \dot{q}_j\right\}\right)$$

is driven by the Lagrangian equation of motion (4.42), the time evolution equation for $S(t)$ becomes the **Hamilton-Jacobi equation**:

$$\frac{\partial S}{\partial t} + H\left(\mathbf{q}, \nabla S\right) = 0, \quad \mathbf{p} = \nabla S. \tag{4.43}$$

In this formalism, the relation $\mathbf{p} = \nabla S$ provides the geometric interpretation that "**the normal to the wavefront** S **determines the trajectory**" (Fig. 4.5). Thus, the time evolution of the wave propagation S is governed by the Hamilton-Jacobi equation. This interpretation is referred to as the **waveform description of mechanics**.

References

1. Mansfield MM, O'Sullivan C (2020) Understanding physics, 3rd edn. Wiley, Hoboken. ISBN: 978-1119519508
2. Riley KF, Hobson MP, Bence SJ (2006) Mathematical methods for physics and engineering, 3rd edn. Cambridge University Press, Cambridge. ISBN: 978-0521679718
3. Wells DA (1967) Schaum's outline of theory and problems of lagrangian dynamics. McGraw-Hill, New York. ISBN: 978-0813232102

4. Born M, Wolf E (2019) Principles of optics (60th Anniversary Edition). Cambridge University Press, Cambridge. ISBN: 978-1108477437
5. Callen HB (1985) Thermodynamics and an introduction to thermostatistics. Wiley, New York. ISBN: 978-0471862567
6. Landau LD, Lifshitz EM (1982) Mechanics. Butterworth-Heinemann, Oxford. ISBN: 978-0750628969
7. Goldstein H, Poole CP, Safko JL (2013) Classical mechanics. Pearson, Harlow. ISBN: 978-1292026558
8. Yamamoto Y, Nakamura K (1998) Kaiseki Rikigaku (in Japanese). Asakura, Tokyo. ISBN: 978-4254136715
9. Bertsekas DP (2005) Dynamic programming and optimal control. Athena Scientific, Nashua. ISBN: 978-1886529083
10. Stengel RF (1994) Optimal control and estimation. Dover, New York. ISBN: 978-0486682006

Chapter 5
Outlined Introduction to Quantum Mechanics

Building upon the understanding from previous chapters, particularly the Hamilton–Jacobi formulation of mechanics, we now proceed to the introduction of quantum mechanics (Sect. 5.1).[1] Quantum mechanics encompasses a vast range of topics, and its content is inherently extensive. However, in this book, we limit the discussion to the minimum necessary for understanding the fundamentals of quantum annealing. Specifically, we focus on two key concepts: superposition states (Sect. 5.2)[2] and tunneling (Sect. 5.3). For instance, topics such as "matrix mechanics leading to the uncertainty principle", which are typically included in elementary courses, have been omitted.

5.1 Optical-Mechanical Analogy and Wave Mechanics

From the late nineteenth to the early twentieth century, it was pointed out that microscopic phenomena involving particles like electrons were in direct contradiction with the predictions of classical physics, leading to the birth of quantum mechanics [2]. One well-known issue was the problem of atomic structure. Experimental investigations established the picture of electrons orbiting a nucleus, but this model led to the conclusion that electrons would radiate energy and spiral into the nucleus, failing to explain the stability of atoms. Another issue was the blackbody radiation problem, where the spectral distribution of thermal radiation could not be explained by classical electromagnetism or thermodynamics. To address these issues, many

[1] This approach is based on the pilot wave concept [1]. While certain deficiencies have been pointed out from a modern perspective in the detailed analysis of quantum mechanics, it provides a logically straightforward and intuitive bridge for beginners to achieve a smooth conceptual landing in the subject.

[2] The meaning of superposition states is further explored in Sect. 7.3.

R. Maezono, *Essentials for Deeper Understanding of Quantum Computing*,
https://doi.org/10.1007/978-981-96-5646-2_5

researchers took different "paths up the mountain" during the same period, gradually uncovering the governing principles of quantum mechanics. One such conceptual path was based on the Hamilton-Jacobi formalism discussed at the end of Sect. 4.6. The breakthrough idea was inspired by the analogy between geometrical optics and wave optics (Sect. 5.1.1). The Hamilton-Jacobi formalism offers a framework for describing wave mechanics, contrasting with the traditional particle-based depiction of mechanics. Using this analogy as a lever, researchers derived the principles of quantum mechanics, which we will outline in this section.

5.1.1 Optics and Mechanics

The propagation of light has long been intuitively understood as light rays. Experimental facts about the propagation of light have been inductively summarized into the laws of rectilinear propagation, reflection, and refraction, forming the framework known as **geometrical optics**. In Sect. 3.4.2, we mentioned that "light rays were identified as the progression of electromagnetic waves ($\mathbf{S} = \mathbf{E} \times \mathbf{B}$)". Indeed, from Maxwell's equations governing electromagnetic waves (Sect. 3.4.1), it can be derived that the Poynting vector **S** reproduces the laws of geometrical optics, such as reflection and refraction [3] (**wave optics**).

The fundamental question of "What is light?" sparked two opposing views. Newton proposed the **particle theory of light**, indparticle theory of light reasoning that light travels in straight lines. In contrast, Huygens supported the **wave theory of light**. Later, diffraction experiments provided evidence for the wave nature of light. With the development of electromagnetism, it became clear that light is an electromagnetic wave. In wave optics, it was shown that taking the zero-wavelength limit reduces wave optics to geometrical optics [3]. This implies that "Light is a wave with wavelength λ, but because the characteristic length scale L of our world is much larger than λ ($\lambda \ll L$), diffraction is usually negligible, and light appears to travel as particle-like rays". This relationship forms the basis of the connection between **wave optics and geometrical optics**.[3]

The laws of geometrical optics can be succinctly summarized by Fermat's principle, which states that the path of a light ray is determined such that the integral

$$I = \int_{P_1}^{P_2} ds \cdot n(\mathbf{r})$$

[3] The Newton-Huygens debate concluded with the idea that "light is a wave", but experimental evidence later suggested that the exchange of energy between light and matter occurs in discrete packets. This insight paved another "path up the mountain" that led to the birth of quantum mechanics. This phenomenon can be understood through the quantization of electromagnetic waves, explained using the scheme known as second quantization, as mentioned in Fig. 9.4.

is minimized.[4] That is, by requiring the minimization of this integral, the laws of straight-line propagation, reflection, and refraction can be derived. It was astonishing to discover that the workings of natural laws governing this world could be encapsulated by such a concise principle. This sparked intellectual curiosity about whether a similar "optical description" might exist for the dynamical laws governing phenomena such as planetary motion. This pursuit was one of the motivations behind the development of analytical mechanics. Just as the various theorems of geometrical optics are unified under Fermat's principle, the various theorems of mechanics are unified under the principle of least action, as we saw in Sect. 4.3.2.

In classical mechanics, a **particle's trajectory** refers to the path determined by the momentum vector $\mathbf{p}$. In the Hamilton-Jacobi formulation of mechanics (Sect. 4.5), as shown in Fig. 4.5a, the momentum vector is given as the normal to a wavefront by $\mathbf{p} = \nabla S$. The time evolution of this "mechanical wavefront S" is described by the Hamilton-Jacobi equation (4.43). However, in classical mechanics, the concept of "a wave S that gives its normal as the trajectory" remains a mathematical convenience, an interpretation introduced for utility rather than a consideration of something real. There was no deeper exploration of S as a physical entity. The motivation for such considerations arose from the abstract interest in exploring what would happen if mechanics were expressed in optical form, similar to the way geometrical optics is derived from wave optics.

5.1.2 Insights from the Optical-Mechanical Analogy

In geometrical optics, the primary object of study is the light ray, which can be interpreted as the trajectory of a photon. Although light is inherently a wave (an electromagnetic wave) with a finite wavelength λ, the relationship $\lambda \ll L$, where L is the characteristic spatial scale, ensures that wave properties like diffraction are negligible. This makes the wave appear as if it follows a particle trajectory.

Figure 4.5b illustrates a conceptual diagram of wave behavior offshore and within a harbor. Offshore, the spatial features are on the scale of kilometers (e.g., bays or peninsulas), corresponding to L. For waves with wavelengths of a few meters, the condition $\lambda \ll L$ applies, resulting in "wavefronts moving in a definite direction". This situation can be equivalently described by the motion of $\mathbf{p}$, defined as the wavefront normal, which corresponds to a particle-like trajectory. However, within the harbor, the spatial features reduce to structures of a few meters, making $\lambda \sim L$. In this regime, diffraction becomes prominent, and the notion of "wavefronts moving in a definite direction" breaks down. Instead, a description of "standing waves in permissible modes" becomes more appropriate:

[4] Here, $n(\mathbf{r})$ represents the spatial distribution of the refractive index, and the integral I is the line integral along the path from point P_1 to P_2 (Sect. 2.5.1).

- **Wavelength** $\ll$ **Spatial Scale**:
  Diffraction is negligible, wavefronts travel straight, and the wavefront normal appears as the trajectory.
- **Wavelength** $\sim$ **Spatial Scale**:
  Diffraction becomes significant, and the wavefront normal ceases to provide a clear trajectory.

This idea, wherein the relative sizes of λ and L dictate whether a particle or wave picture is appropriate, provided a crucial insight for quantum mechanics. The problem of "why atoms can exist in stable", a mystery of the time, suggested that the spatial scale of atoms is comparable to the wavelength λ, rendering the particle description of "an electron orbiting the nucleus" inappropriate. Electrons, discovered in particle beam experiments such as Crookes tubes [2], were initially identified as particles. However, electrons may truly be waves with wavelength λ. In a vacuum with no surrounding structures, they resemble "waves offshore" in Fig. 4.5b, where diffraction is negligible, and the wavefront normal, $\mathbf{p} = \nabla S$, appears clearly particle-like. This leads to the idea that the equation describing such a wavefront S is none other than the Hamilton-Jacobi equation:

$$\frac{\partial S}{\partial t} + H\left(\mathbf{q}, \nabla S\right) = 0, \quad \mathbf{p} = \nabla S \tag{5.1}$$

5.1.3 Wave Mechanics

In classical mechanics, the mechanical wave S was treated merely as a "mathematical convenience in analogy", rather than being considered a physical entity. However, through the considerations in the preceding subsection, a new perspective emerges (Fig. 5.1): what if the appearance of particles is merely a special limit, and the wave nature represents the true physical reality? This prompts a closer and more serious consideration of Eq. (5.1).

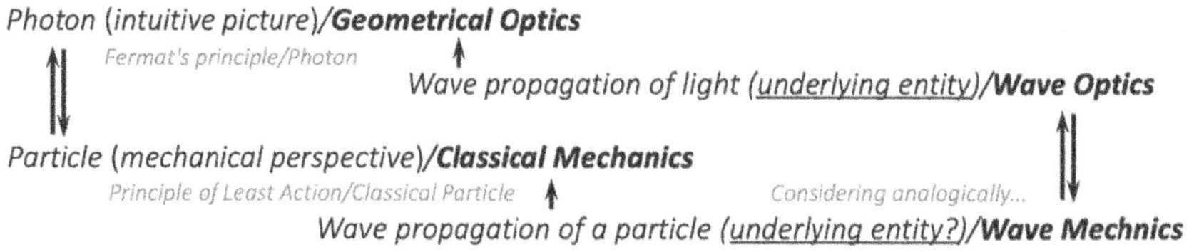

Fig. 5.1 The correspondence between optics and mechanics: In analogy with the relationship between geometrical optics and wave optics, classical mechanics and wave mechanics are juxtaposed

S represents a wavefront, or a phase surface, and the corresponding wave ϕ takes the form $\phi \sim \exp[i \cdot S]$ (Sect. 2.4.9). Namely, the "mechanical wave" ϕ can be expressed as:

$$\phi(\mathbf{r}, t) = A \cdot \exp[i\alpha \cdot S(\mathbf{r}, t)], \quad \therefore \quad i\alpha \cdot S = \ln\phi - \ln A, \tag{5.2}$$

where A and α are constants. The time derivative of S that appears in the Hamiltonian-Jacobi equation is therefore:

$$i\alpha\frac{\partial S}{\partial t} = \frac{1}{\phi}\frac{\partial\phi}{\partial t}.$$

Substituting this into the Hamiltonian-Jacobi equation gives:

$$\frac{1}{i\alpha}\frac{1}{\phi}\frac{\partial\phi}{\partial t} + H = 0, \quad \therefore \quad \frac{i}{\alpha}\frac{\partial\phi}{\partial t} = H\phi. \tag{5.3}$$

Here,

$$\phi(\mathbf{r}, t) = \chi(t)\,\psi(\mathbf{r})$$

is separated into variables (Sect. 2.4.6), and assuming that the Hamiltonian has no explicit time dependence,

$$\frac{i}{\alpha}\frac{\partial\chi(t)}{\partial t}\psi(\mathbf{r}) = \chi(t)\,H \cdot \psi(\mathbf{r})$$

$$\therefore \quad \frac{i}{\alpha\chi(t)}\frac{\partial\chi(t)}{\partial t} = \frac{H \cdot \psi(\mathbf{r})}{\psi(\mathbf{r})} = E$$

is obtained. Recalling that the Hamiltonian has the same dimension as energy, it is evident that the constant E has the dimension of energy. Now, the separated simultaneous equations are:

$$H \cdot \psi(\mathbf{r}) = E \cdot \psi(\mathbf{r}) \tag{5.4}$$

$$\frac{\partial\chi(t)}{\partial t} = -iE\alpha\chi(t). \tag{5.5}$$

The latter equation concerning time can be solved as:

$$\chi(t) = \exp[-iE\alpha \cdot t]. \tag{5.6}$$

Regarding Eq. (5.4) for the remaining spatial part, at this point in the discussion, H is merely a "function", so it is unclear how to treat:

$$H \cdot \psi(\mathbf{r}) = E \cdot \psi(\mathbf{r}). \tag{5.7}$$

Here, we make the following leap. From the framework of the Hamilton-Jacobi method, $\mathbf{p} = \nabla S$, using Eq. (5.2) for evaluation gives:

$$\mathbf{p} = \nabla S = \frac{1}{i\alpha}\frac{1}{\psi}\nabla\psi, \quad \therefore \quad \mathbf{p}\psi = \frac{\nabla}{i\alpha}\psi.$$

This can be seen as:

$$\hat{\mathbf{p}} = \frac{\nabla}{i\alpha}, \tag{5.8}$$

indicating that "momentum $\mathbf{p}$ can be reinterpreted as the operator $\hat{\mathbf{p}}$". On the other hand, from Eq. (5.5),

$$E = -\frac{1}{i\alpha}\frac{\partial}{\partial t},$$

can formally be written. These lead to the quantization rules, known as the **correspondence principle**:

$$\begin{pmatrix} \mathbf{p} \\ E \end{pmatrix} \to \frac{1}{i\alpha}\begin{pmatrix} \nabla \\ -\partial_t \end{pmatrix}, \quad \left[\alpha = 1/\hbar \text{ as identified later (explained later)}\right]. \tag{5.9}$$

Applying this to the Hamiltonian:

$$H = \frac{p^2}{2m} + U(\mathbf{r}),$$

it transforms into the operator:

$$\hat{H} = -\frac{\nabla^2}{2m\alpha^2} + U(\mathbf{r}).$$

Consequently, Eq. (5.7) for the spatial part becomes:

$$\left(-\frac{\nabla^2}{2m\alpha^2} + U(\mathbf{r})\right)\cdot\psi(\mathbf{r}) = E\cdot\psi(\mathbf{r}), \tag{5.10}$$

which can be interpreted as an eigenvalue problem described by a differential equation (Sect. 2.4.3). This equation is called the **Schrodinger equation**.

In this equation, by setting the "Coulomb potential from a nucleus placed at the origin", $U(\mathbf{r}) = Z/|\mathbf{r}|$, it represents the problem of an isolated atom with atomic number Z. By solving the eigenvalue problem of this differential equation, the eigenvalues E are mathematically obtained as discrete values: $E = E_0, E_1, E_2, \cdots$ (Sect. 2.4.3). These discrete values are derived as $E_n \sim 1/n^2$. It is shown that this pattern of discrete intervals corresponds to the experimentally known discrete frequencies ω_n of the emission spectrum from atoms[5] at the time [2, 4]. From Eq. (5.6), the energy eigenvalue E indeed corresponds to the frequency ω. By setting the constant α as $\alpha = 1/\hbar$, it matches the experimental values. Hence, the constant previously denoted as α is identified as the reciprocal of Planck's constant $\hbar = 1.05\cdots \times 10^{-34}\,\mathrm{J} \cdot \mathrm{s}$, originally discovered in the context of the blackbody radiation problem (Sect. 5.1.1).

5.2 Physical Quantities and Observations in Quantum Mechanics

5.2.1 *Operators and Observations*

Reflecting $\alpha = 1/\hbar$, it is restated that in the regions where quantum phenomena become prominent, characterized by scales comparable to the wavelength of mechanical waves where interference is significant, the following relationships hold:

$$\phi(\mathbf{r}, t) \sim \exp\left[i \cdot \frac{S(\mathbf{r}, t)}{\hbar}\right], \quad \mathbf{p}(\mathbf{r}, t) = \nabla S(\mathbf{r}, t) . \tag{5.11}$$

The function $\phi(\mathbf{r}, t)$ is called the **wave function**. The time evolution of the wave function, based on Eq. (5.3), is expressed as:

$$i\hbar \frac{\partial \phi(\mathbf{r}, t)}{\partial t} = H\phi(\mathbf{r}, t) . \tag{5.12}$$

Rewriting Eq. (5.8):

$$\hat{\mathbf{p}} = \frac{\hbar}{i}\nabla , \tag{5.13}$$

indicating that **momentum corresponds to an operator**. When the spatial part ψ of the wave function $\phi(\mathbf{r}, t) \sim \psi(\mathbf{r})$ takes the form of a simple plane wave $\psi \sim \exp(i\mathbf{k}\mathbf{r})$,

$$\hat{\mathbf{p}} \cdot \psi = \frac{\hbar}{i}\nabla\psi = \hbar\mathbf{k} \cdot \psi , \tag{5.14}$$

which reproduces the relationship $\mathbf{p} = \hbar\mathbf{k}$, historically suggested by experiments such as Compton scattering [2]. This operation of "applying the operator to ψ" can

[5] This refers to the series of experimental facts summarized by Rydberg's formula, including the Lyman series, Balmer series, Paschen series, and Brackett series.

be interpreted as "applying the operator $\hat{\mathbf{p}}$ to the state ψ and obtaining $\hbar\mathbf{k}$ as the result of the **observation**".

In the Hamiltonian formalism (Sect. 4.4.3), physical quantities are expressed as $F = F(\mathbf{q}, \mathbf{p})$, with coordinates $\mathbf{q}$ and momentum $\mathbf{p}$ treated as independent variables. In the current framework of wave mechanics, this implies that general physical quantities transform as:

$$F(\mathbf{q}, \mathbf{p}) \to \hat{F}\left(\mathbf{q}, \frac{\hbar}{i}\nabla\right),$$

where, the momentum part is replaced by an operator. This is exemplified by the Schrodinger equation stated in (5.10):

$$\hat{H}\left(\mathbf{q}, \frac{\hbar}{i}\nabla\right) \cdot \psi(\mathbf{r}) = E \cdot \psi(\mathbf{r}).$$

More generally, it suggests the relationship:

$$\hat{F} \cdot \psi = F \cdot \psi,$$

where physical quantities are determined through the procedural depiction of an "operator action (observation) $\hat{F}$" resulting in an "eigenvalue F" as the response. What was understood as physical quantities in classical theory has been separated into three entities in quantum theory: the observable, the operator, and the state. These entities are related by the principle that an operator acting on a state yields the observable.

5.2.2 *Expectation Values of Observables*

Recalling the Fourier expansion discussed in Sect. 2.4.5, the general functional form of the spatial part ψ can be expressed as a Fourier expansion containing various wave number components:

$$\psi = c_{\mathbf{k}_1} \cdot \exp[i\mathbf{k}_1\mathbf{r}] + c_{\mathbf{k}_2} \cdot \exp[i\mathbf{k}_2\mathbf{r}] + \cdots .$$

Using the bra-ket notation introduced in Sect. 2.4.4, this is written as:

$$|\psi\rangle = c_{\mathbf{k}_1} \cdot |\mathbf{k}_1\rangle + c_{\mathbf{k}_2} \cdot |\mathbf{k}_2\rangle + \cdots . \tag{5.15}$$

Each expansion basis satisfies:

$$\hat{\mathbf{p}} \cdot |\mathbf{k}_j\rangle = \hbar\mathbf{k}_j \cdot |\mathbf{k}_j\rangle,$$

which indicates that these are eigenvectors, and $|\psi\rangle$ corresponds to a general vector expanded in terms of eigenvectors. In the "single-color" case described by Eq. (5.14), where $c_{\mathbf{k}_j} = \delta_{jm}$ (all components vanish except for a single $\mathbf{k}_m$), this state corresponds to situations like "waves offshore" in Fig. 4.5b, where "classical trajectories and traveling waves align clearly" (recovering the classical picture). Conversely, situations like "waves within a harbor" in the same figure, where diffraction causes scattering in various directions, correspond to a state with mixed components as in Eq. (5.15). In this case:

$$\hat{\mathbf{p}} \cdot |\psi\rangle = \sum_{\mathbf{k}} c_{\mathbf{k}} \cdot \hbar\mathbf{k} \cdot |\mathbf{k}\rangle \neq \lambda \cdot |\psi\rangle,$$

which shows that $\hat{\mathbf{p}} \cdot |\psi\rangle$ is not proportional to $|\psi\rangle$ and transforms into a different state. This leads to the question: how should we interpret a state like Eq. (5.15) within the framework of "what is returned when an observable is measured on the state $|\psi\rangle$," as described in Eq. (5.14)?

To generalize the discussion, consider a physical quantity $\hat{F}$ and its eigenstates satisfying

$$\hat{F} \cdot |\chi_m\rangle = F_m \cdot |\chi_m\rangle,$$

where the mechanical wave is expanded as

$$|\psi\rangle = c_1|\chi_1\rangle + c_2|\chi_2\rangle + \cdots . \tag{5.16}$$

We consider the meaning of this state. Applying the operator $\hat{F}$, we obtain

$$\hat{F}|\psi\rangle = F_1 \cdot c_1 \cdot |\chi_1\rangle + F_2 \cdot c_2 \cdot |\chi_2\rangle + \cdots .$$

Here, each term on the right-hand side represents a "monochromatic state", interpreted as "if observed, the state $|\chi_j\rangle$ is exclusively realized, yielding the observable value F_j." Taking the inner product with $\langle\psi|$ on both sides, and using the orthogonality condition $\langle\chi_i|\chi_j\rangle \sim \delta_{ij}$,[6] we obtain

$$\langle\psi|\hat{F}|\psi\rangle = |c_1|^2 \cdot F_1 + |c_2|^2 \cdot F_2 + \cdots = \langle F\rangle .$$

In the rightmost term, $\langle F\rangle$ is written because this equation can be interpreted as the "expectation value $\langle F\rangle$ weighted by $|c_j|^2$". Since the occurrence of each state is exclusive, the state described by Eq. (5.16) can thus be interpreted as a "state (**quantum superposition state**) in which each state occurs with a probability $|c_j|^2$ upon observation".

---

[6] This can be proven for real eigenvalues of Hermitian operators [5].

Since $\left|c_j\right|^2$ is interpreted as the probability of occurrence,

$$\langle\psi|\psi\rangle = |c_1|^2 + |c_2|^2 + \cdots$$

must be normalized to 1. Therefore, this normalization factor is sometimes explicitly included as:

$$F = \frac{\langle\psi|\hat{F}|\psi\rangle}{\langle\psi|\psi\rangle} = |c_1|^2 \cdot F_1 + |c_2|^2 \cdot F_2 + \cdots .$$

In the above reasoning using the expansion in terms of eigenfunctions, note that the **linearity of the governing equation** described in Sect. 2.4.1 is an essential assumption. In other words, "each exclusive possibility $|\chi_j\rangle$", which individually serves as a solution to the governing equation (eigenstate), also implies that their superposition (mixed state) is likewise a solution.

Where Exactly Are Beginners Struggling?

Beginners often find themselves confused not at the advanced stages of quantum mechanics but rather at the very foundation of its mathematical framework. Specifically, they struggle with the concept that "any general vector can be expanded in terms of eigenvector systems", as presented in Eq. (5.16). This fundamental principle of linear algebra is introduced abruptly, often without sufficient preparation or understanding, in early quantum mechanics courses. In fields like material science, first-year students are required to learn quantum mechanics quickly to progress to more specialized courses involving electronic theories. This rapid introduction often comes before they have developed a strong grasp of linear algebra, leaving many feeling lost. The actual issue lies in the disconnect between the foundational mathematical concepts and their application in quantum mechanics. While instructors might assume that students are grappling with the philosophical or interpretative challenges of quantum mechanics (e.g., measurement interpretation), the majority are instead struggling with the basic mechanics of eigenfunction expansion and what it means in practice.

5.2.3 Quantum Mechanical Superposition

Let us consider the meaning of expressing the mechanical wave as $\phi(\mathbf{r}, t) \sim \exp[i \cdot S/\hbar]$. This expression implies that the "length scale over which the equiphase surface S changes by approximately $\hbar$" corresponds to the wavelength of the mechanical wave $\phi(\mathbf{r}, t)$.[7] The Planck constant, $\hbar = 1.05 \times 10^{-34}$ J·s, is an exceedingly small quantity. The spatial variation $\Delta r \sim \lambda$, which induces such a minuscule change in S, also becomes exceedingly small as a length scale. When the characteristic scale L of the space through which the mechanical wave propagates approaches such an extremely small wavelength λ, diffraction of the mechanical wave becomes pronounced, and the problem requires "quantum mechanical description" (see Fig. 4.5).

In our daily lives, such situations rarely occur, allowing classical descriptions like "observing the trajectory of a particle as it appears" to hold true. However, in microscopic scenarios, such as those involving atoms, the domain shifts to one where "quantum mechanical descriptions are necessary". In this domain, the mathematics of "operators/observables" governs, and superposition states like those expressed in Eq. (5.16) are prominently recognized as representations of physical states. As widely discussed in many popular science books, these superposition states lead to outcomes incompatible with classical picture of our's daily intuition.

The superposition state described by (5.16) is interpreted as a state in which "multiple mutually exclusive states $|\chi_1\rangle, |\chi_2\rangle, \cdots$ coexist and evolve simultaneously, and upon measurement, one of these states appears with probability $|c_m|^2$" (the **Born interpretation**).[8] The predictions based on the Born interpretation, have indeed matched experimental facts. As a result, this interpretation has been reluctantly accepted over time, despite its elusive nature. This interpretative aspect represents the core of the statement, "the more you think about it, the less clear it becomes", or "no one truly understands it". Thought experiments derived from this interpretation, such as the famous "Schrodinger's cat" and the double-slit experiment [7], have highlighted peculiar conclusions. In the double-slit thought experiment, a single particle simultaneously realizes two possibilities: "passing through slit A" and "passing through slit B". The particle generates interference effects between these two possibilities and determines its landing point while experiencing the results of this interference [7]. In modern times, this outcome has been directly confirmed through experiments [8]. As such, the modern approach to quantum mechanics is to accept these "interpretations" as "axioms", starting the discussion from this premise. This represents the contemporary style of quantum mechanics textbooks [9].

---

[7] Recall the representation of the phase angle as $\exp[i \cdot \theta(\mathbf{r})]$ from Sect. 2.4.9. The phase change of $\theta(\mathbf{r})$ by $2\pi \sim O(1)$ was equivalent to "one complete cycle" of the wave. Currently, with $\theta = S/\hbar$, a change of $\theta \sim O(1)$ corresponds to a change in $S \sim \hbar$.

[8] This is not an issue of the mathematical representation itself, but rather a matter of "what the mathematical representation describes". During the period when the authors received their education, the Born interpretation was the dominant interpretation. However, there have been various other interpretations [6], and these interpretations later stimulated the rise of the quantum computing field.

A quintessential example of the superposition state described by Eq. (5.16) is the two-state system, known as a **qubit**, expressed as

$$|\psi\rangle = c_0 \cdot |0\rangle + c_1 \cdot |1\rangle \tag{5.17}$$

In this system, the mutually exclusive states $|0\rangle$ and $|1\rangle$ are not exclusively realized (as in classical bits in digital computing), but instead, the framework allows for "simultaneous parallel evolution of mutually exclusive states until observed". This framework enables a fundamentally different approach compared to classical computing, which processes mutually exclusive possibilities sequentially. Instead, **quantum computing** allows for "simultaneous processing of mutually exclusive possibilities by maintaining their coexistence in parallel".

5.3 Path Integral Formulation

Combinatorial optimization problems involve "finding the lowest valley among many local valleys" in a multi-peaked potential. Quantum annealing is particularly effective for tackling combinatorial optimization problems that are otherwise infeasible for classical computation. The reasoning is explained as follows: "In classical mechanics, energy barriers cannot be crossed, whereas in quantum mechanics, such barriers can be traversed via **virtual transitions**". Alternatively, it is described as "quantum mechanics allows **tunneling** through energy barriers". This ability to tunnel through or bypass multi-peaked potential barriers enables finding the lowest solution. However, the "key concept" that "quantum mechanics overcomes energy barriers via virtual transitions" appears abruptly as an axiom, leaving beginners confused and perplexed. The path integral formulation, discussed in this section, provides an intuitive framework to understand this phenomenon.

In the optical-mechanical analogy, the mechanical wave ϕ associated with the action integral S and level surfaces of S is given as $\phi \sim \exp\left[iS/\hbar\right]$. Consequently, the quantum mechanical dynamics describing the evolution of $\phi(x_0, t_0)$ to $\phi(x, t)$ can be expressed as:

$$\phi(x,t) \sim \exp\left[\frac{i}{\hbar} S(x, x_0)\right] \cdot \phi(x_0, t_0),$$

as shown in Fig. 5.2. By summing over all possible connection points x_0, this expression generalizes to:

$$\phi(x,t) \sim \int_{x_0} dx_0 \cdot \exp\left[\frac{i}{\hbar} S(x, x_0)\right] \cdot \phi(x_0, t_0). \tag{5.18}$$

In the following, we derive that this expression can indeed be written in the form of Eq. (5.21).

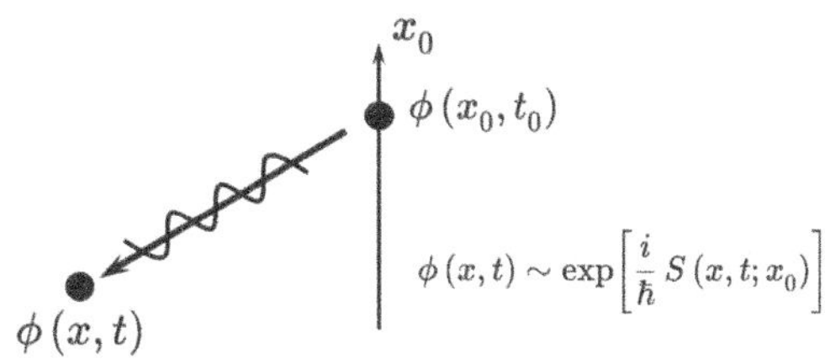

Fig. 5.2 Propagation form of the wave function. At the initial point $\phi(x_0, t_0)$, the phase $\exp[iS(x,t;x_0)/\hbar]$ acquired during the transition $(x_0, t_0) \rightarrow (x, t)$ is applied, resulting in $\phi(x, t)$

5.3.1 Dirac's Derivation

Using the bracket notation $\phi(x,t) = \langle x|\phi(t)\rangle$, the time evolution of Eq. (5.12) can be written as:

$$i\hbar\frac{\partial}{\partial t}|\phi(t)\rangle = \hat{H}|\phi(t)\rangle.$$

This can be formally solved as:

$$|\phi(t)\rangle = \exp\left[-\frac{i}{\hbar}\hat{H}\cdot t\right]|\phi(0)\rangle.$$

This can also be expressed as:

$$|\phi(t)\rangle = U(t,t_0)\cdot|\phi(t_0)\rangle, \quad U(t-t_0) = \exp\left[-\frac{i}{\hbar}\hat{H}\cdot(t-t_0)\right]. \tag{5.19}$$

Applying the above equation to $\phi(x,t) = \langle x|\phi(t)\rangle$, and inserting the identity relation for x_0 [Eq. (9.26)], we obtain:

$$\langle x|\phi(t)\rangle = \langle x|U(t,t_0)|\phi(t_0)\rangle = \langle x|U(t,t_0)\underline{\int dx_0|x_0\rangle\langle x_0|}\cdot|\phi(t_0)\rangle$$
$$= \int dx_0\,\langle x|U(t,t_0)|x_0\rangle\,\langle x_0|\phi(t_0)\rangle\,.$$

Rewriting this, we have:

$$\phi(x,t) = \int dx_0\,\langle x|U(t,t_0)|x_0\rangle\,\phi(x_0,t_0) =: \int dx_0\cdot K(x,x_0)\,\phi(x_0,t_0)\,,$$

which can be expressed in a **propagation form**. The **propagation kernel**:

$$K(x, x_0) = \langle x|U(t, t_0)|x_0\rangle$$

is referred to as the **Feynman kernel**. Evaluating this Feynman kernel leads to a form like Eq. (5.18), which we will derive below.

Hereafter, to simplify notation, we set $\hbar = 1$.[9] Dividing $\Delta t = N \cdot \varepsilon$, we express

$$K(x, x_0) = \langle x|e^{-i\hat{H}\cdot\Delta t}|x_0\rangle = \langle x|e^{-i\hat{H}\cdot\varepsilon} \cdot e^{-i\hat{H}\cdot\varepsilon} \cdots e^{-i\hat{H}\cdot\varepsilon}|x_0\rangle\,.$$

By inserting the identity relation [(Eq. 9.26)]

$$1 = \int dx_j \cdot |x_j\rangle\langle x_j|,$$

at intermediate steps, we obtain

$$K(x, x_0) = \int dx_N dx_{N-1} \cdots dx_1 \cdot \left\langle x|e^{-i\hat{H}\cdot\varepsilon}|x_N\right\rangle\left\langle x_N|e^{-i\hat{H}\cdot\varepsilon}|x_{N-1}\right\rangle \times \cdots \times \left\langle x_1|e^{-i\hat{H}\cdot\varepsilon}|x_0\right\rangle. \tag{5.20}$$

For the exponential term $e^{-i\hat{H}\cdot\varepsilon} = e^{-i\varepsilon\left(\hat{T}+V\right)}$, note that $\hat{T}$ and V are non-commutative, so $e^{-i\left(\hat{T}+V\right)\cdot\varepsilon} \neq e^{-i\hat{T}\cdot\varepsilon} \cdot e^{-iV\cdot\varepsilon}$. This part is handled using the Suzuki-Trotter formula [10],

$$e^{-i\hat{H}\cdot\varepsilon} = e^{-i\varepsilon\left(\hat{T}+V\right)} = e^{-i\varepsilon V/2} \cdot e^{-i\varepsilon\hat{T}} \cdot e^{-i\varepsilon V/2} + O\left(\varepsilon^3\right).$$

If ε is sufficiently small, $O\left(\varepsilon^3\right)$ can be neglected. By inserting the identity relation over p, we have[10]

$$\begin{aligned}\left\langle x_k|e^{-i\hat{H}\cdot\varepsilon}|x_{k-1}\right\rangle &= \left\langle x_k|e^{-iV\cdot\varepsilon/2}e^{-i\hat{T}\cdot\varepsilon} \cdot e^{-iV\cdot\varepsilon/2}|x_{k-1}\right\rangle \\ &= \underline{\int dp_k}\left\langle x_k|e^{-iV\cdot\varepsilon/2}e^{-i\hat{T}\cdot\varepsilon}\underline{|p_k\rangle\langle p_k|}e^{-iV\cdot\varepsilon/2}|x_{k-1}\right\rangle\end{aligned}$$

---

[9] This corresponds to substituting $H/\hbar \to H$, meaning energy is measured in units of "multiples of $\hbar$". Similarly, conventions such as setting "the speed of light as $c = 1$" or "the electron mass as $m_e = 1$" might confuse beginners. However, understanding that this simply counts quantities in terms of "multiples of the set unit" should alleviate much of the confusion.

[10] The terms involving V, which is a c-number expressed in x, interact with the nearest neighbors, specifically the "left-side $\langle x_k|$ or right-side $|x_{k-1}\rangle$", resulting in factors like $e^{-iV(x_{k-1})\cdot\varepsilon/2}$. Meanwhile, the operator $\hat{T}$, expressed in p, acts on the nearest right-side $|p_k\rangle$, extracting the value of p_k.

$$= \int dp_k \cdot e^{-iV(x_k)\cdot\varepsilon/2} \exp\left[-i\frac{p_k^2}{2m}\cdot\varepsilon\right]$$

$$e^{-iV(x_{k-1})\cdot\varepsilon/2}\langle x_k|p_k\rangle\langle p_k|x_{k-1}\rangle .$$

By applying the results of (9.27) and (9.28),

$$\langle x_k|p_k\rangle = \frac{1}{\sqrt{2\pi}}\cdot e^{ip_k x_k}, \quad \langle p_k|x_{k-1}\rangle = \frac{1}{\sqrt{2\pi}}\cdot e^{-ip_k x_{k-1}},$$

we can evaluate[11]

$$\left\langle x_k|e^{-iH\cdot\varepsilon}|x_{k-1}\right\rangle = \frac{1}{2\pi}\int dp_k\cdot\exp\left[-i\left\{\frac{p_k^2}{2m}+\frac{V(x_k)+V(x_{k-1})}{2}\right\}\cdot\varepsilon\right]$$

$$\cdot e^{ip_k x_k}\cdot e^{-ip_k x_{k-1}}$$

$$\approx \frac{1}{2\pi}\int dp_k\cdot\exp\left[i\left\{p_k\frac{x_k-x_{k-1}}{\varepsilon}-H(x_k)\right\}\cdot\varepsilon\right]$$

$$= \frac{1}{2\pi}\int dp_k\cdot\exp\left[i\left\{p_k\dot{x}_k-H(x_k)\right\}\cdot\varepsilon\right].$$

Substituting this into Eq. (5.20), we obtain

$$K(x,x_0) = \left(\frac{1}{2\pi}\right)^N \int \mathcal{D}x\mathcal{D}p\cdot\exp\left[i\sum_{k=1}^{N+1}\left\{p_k\dot{x}_k-H(x_k)\right\}\cdot\varepsilon\right].$$

Here, we define

$$\mathcal{D}x = dx_1\cdots dx_N, \quad \mathcal{D}p = dp_1\cdots dp_N,$$

which implies the continuous limit as $N\to\infty$ for the discretization $\Delta t = N\cdot\varepsilon$. In this limit, the indices k that appeared due to the discretization of t transition back to continuous variables, $p_k\to p(t)$ and $x_k\to x(t)$. The term $\{p_k\dot{x}_k - H(x_k)\}$ becomes

$$p(t)\dot{x}(t) - H(t) = L(t),$$

[11] Here, the approximation $\left[\frac{p_k^2}{2m}+\frac{V(x_k)+V(x_{k-1})}{2}\right] \approx \left[\frac{p_k^2}{2m}+V(x_k)\right] = H(x_k)$is employed.

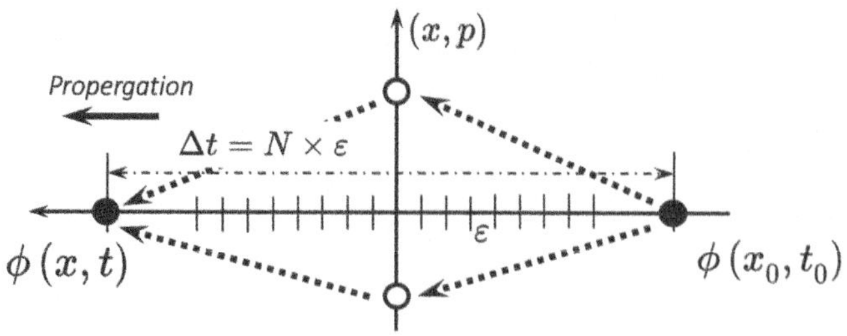

Fig. 5.3 A conceptual diagram illustrating the path integral described by Eq. (5.21). The horizontal axis is segmented into numerous divisions of width ε. At each division, an "integration along the vertical axis" is performed. Although only two paths connected by straight lines are depicted in the figure, the paths contributing to the path integral $\mathcal{D}x\mathcal{D}p$ comprise all possible trajectories extending from the starting point (x_0, t_0) to the endpoint (x, t)

which can be expressed in terms of the Lagrangian L [Eq. (4.29)]. Consequently,

$$K(x, x_0) = \left(\frac{1}{2\pi}\right)^N \int \mathcal{D}x\mathcal{D}p \cdot \exp\left[i \int_{t_0}^{t} L(t') \cdot dt'\right]$$
$$= \left(\frac{1}{2\pi}\right)^N \int \mathcal{D}x\mathcal{D}p \cdot \exp\left[iS(t, t_0)\right].$$

Here, S represents the action integral, as introduced in Eq. (4.26).

Taking into account the earlier replacement $H/\hbar \rightarrow H$ and the dimensional correspondence $H \sim L$, we now reintroduce $\hbar$ into the expression. This yields:

$$\phi(x, t) = \int dx_0 \cdot K(x, x_0)\, \phi(x_0, t_0)$$
$$K(x, x_0) = \left(\frac{1}{2\pi}\right)^N \int \mathcal{D}x\mathcal{D}p \cdot \exp\left[iS(t, t_0)/\hbar\right]. \quad (5.21)$$

Indeed, as stated at the beginning of this section, we have derived the form where "the wave propagates by accumulating the phase $\exp[iS(t, t_0)]$ of the wave across all possible intermediate paths in the phase space (x, p)" (Fig. 5.3).

5.3.2 Correspondence with Classical Mechanics

As established in the previous subsection, quantum mechanical dynamics involves the propagation of states by summing up phase contributions over all possible paths in phase space (x, p) that connect (x_0, t_0) to (x, t). The phase of these waves is

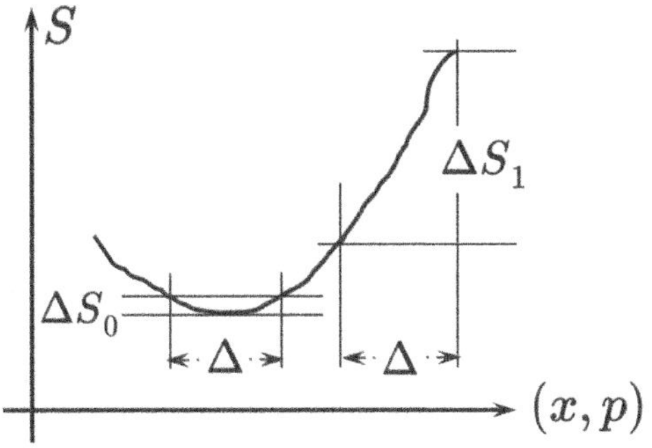

Fig. 5.4 When the same variation Δ is applied in phase space, the change in S near the minimum point, ΔS_0, is small. In contrast, the change ΔS_1 farther from the minimum point becomes larger compared to ΔS_0

$\exp[iS/\hbar]$, and as noted at the beginning of Sect. 5.2.3, a change in S of the order of $\hbar$ corresponds to one wavelength. Additionally, the situation where the characteristic scale of phenomena is much larger than the wavelength $\lambda \sim \hbar$ corresponds to the classical mechanics limit. When $\hbar \to 0$, the characteristic scale of all phenomena always becomes much larger than the wavelength $\lambda \sim \hbar$. Thus, $\hbar \to 0$ describes the **classical mechanics limit**. In the $\hbar \to 0$ limit, even a small variation in $S(t, t_0)$ leads to significant oscillations in the phase angle $\theta = S/\hbar$, causing it to fluctuate extensively. This results in "hitting almost all angles randomly".[12]

Expressing quantum dynamics in the language of path integrals reveals that contributions from possible paths through various points in phase space, represented by the "vertical axis (p, x)" in Fig. 5.3, propagate as waves weighted by the phase factor $\exp[iS/\hbar]$ along each path [(5.21)]. In the classical mechanics limit $\hbar \to 0$, the contributions from each phase factor end up "hitting almost all angles randomly", causing them to cancel each other out and result in zero net contribution.[13] The contributions that barely remain, as explained below, are those around the "minimum point of the action integral" S:[14] For a change ΔS in S, the fact that $\Delta S/\hbar$ oscillates significantly was the reason why contributions cancel each other out to yield zero. As shown in Fig. 5.4, consider the change in S resulting from the same variation Δ in phase space. The change ΔS_0 near the minimum point is smaller compared to the change ΔS_1 further away from the minimum point. Thus,

[12] Imagine a situation where θ moves across the general angular axis in steps such as $\Delta\theta = +1000$ or $-10{,}000$. This helps in understanding a scenario where θ spans all general angles corresponding to $\theta = [-\pi, \pi]$.

[13] This kind of reasoning frequently appears in discussions of quantum algorithms and similar topics.

[14] To be precise, the term "stationary point" should be used. A stationary point refers to a concept analogous to the "minimum" of a function but applied to a functional. The reasoning below applies not only to minimum points but also to stationary points in general. Thus, strictly speaking, the term "stationary point" should be used. However, for the sake of making the concept more accessible to beginners, the term "minimum point" is employed here. That said, stationary points are not necessarily minimum points. Therefore, it is important to become accustomed to using the term "stationary point" at an early stage.

contributions from regions near the minimum point exhibit smaller oscillations compared to those from other regions. In the $\hbar \to 0$ limit, this distinction becomes more pronounced, making contributions near the minimum point the "last remaining contributions" as others cancel out.

From the above discussion, in the classical mechanics limit $\hbar \to 0$, it becomes evident that "only the paths corresponding to the vicinity of the minimum point of the action integral S contribute to the propagation in Eq. (5.21), with 'weights that fail to cancel out'". Recalling the "minimum point of the action integral S", let us relate it to the "principle of least action" described in Sect. 4.3.2. There, it was shown that "the trajectory that minimizes the action integral S is realized as the 'dynamics governed by the classical equation of motion'". Thus, in the classical mechanics limit $\hbar \to 0$ of quantum mechanics expressed in the path integral formulation, we can indeed confirm that "only the trajectories that minimize the action integral S contribute", which corresponds to the return to classical mechanics.

Starting from this classical limit, it can be inferred that in quantum dynamics, the system also evolves over "paths (processes) that do not minimize the action integral S", which are impossible in classical mechanics. This implies that the dynamics proceed through a wide range of possible transitions, including those "state transitions that cannot occur in classical theory". Such transitions are referred to as **virtual transitions**. In classical theory, a system cannot surpass "high energy barriers" that exceed the energy it possesses in its initial state. This corresponds to the fact that "trajectory that minimizes the action integral S" is only realized in the classical mechanics. However, when entering quantum-mechanical regions,[15] state evolution becomes possible through processes such as "temporarily borrowing energy to surpass a high energy barrier, transitioning to another lower energy state, and then repaying the borrowed energy to balance the account".[16]

References

1. Duerr D, Teufel S (2009) Bohmian mechanics. Springer, Berlin. ISBN: 978-3540893431
2. Mansfield MM, O'Sullivan C (2020) Understanding physics, 3rd edn. Wiley, Hoboken. ISBN: 978-1119519508
3. Born M, Wolf E (2019) Principles of optics (60th Anniversary Edition). Cambridge University Press, Cambridge. ISBN: 978-1108477437
4. McQuarrie DA, Simon JD (2023) Physical chemistry. University Science Books. ISBN: 978-1940380216
5. Landau LD, Lifshits EM (1965) Quantum mechanics. Pergamon Press, New York. ISBN-10: 978-0080091013
6. Weinberg S (2015) Lectures on quantum mechanics, 2nd edn. Cambridge University Press, Cambridge. ISBN: 978-1107111660

[15] Regions where diffraction of the wave function becomes prominent.

[16] Yukawa's meson theory was also inspired by the idea of "temporarily adopting a high-energy state…".

7. Feynman RP (2011) The Feynman lectures on physics, New Millennium Edition. Basic Books, New York. ISBN: 978-0465023820
8. Piela L (2013) Ideas of quantum chemistry, 2nd edn. Elsevier, Amsterdam. ISBN: 978-0444594365
9. Nielsen MA, Chuang IL (2013) Quantum computation and quantum information. Cambridge University Press, Cambridge. ISBN: 978-1107619197
10. Gould H, Tobochnik J, Christian W (2017) An introduction to computer simulation methods: applications to physical systems, 3rd edn. Amazon Digital Services, Seattle. ISBN: 978-1974427475

Chapter 6
Overview of Relativistic Theory

In this chapter, the goal is to understand how the Pauli matrices $\boldsymbol{\sigma}$, which appear in spin descriptions, couple with the magnetic field **B** through an inner product ($H \sim \boldsymbol{\sigma} \cdot \mathbf{B}$) (Sect. 6.3.2).[1] This outcome can be understood as a consequence of evaluating the "fundamental equation of relativistic electron theory (Dirac equation)" in the non-relativistic limit (Sect. 6.3.1). A series of explanations leading to this goal will be systematically developed. First, in Sect. 6.2, as part of the "Overview of Relativistic Mechanics", the relation $E^2 = c^2p^2 + m^2c^4$ is derived, and the correspondence principle (5.9) is applied to derive the Dirac equation. To ensure a smooth understanding of the "Overview of Relativistic Mechanics", the discussion begins with a conceptual review of "Vectors and Scalars" (Sect. 6.2.1) and proceeds to extend the concept of canonical forms involving the Lagrangian and Hamiltonian (Sect. 6.1).

6.1 Generalization of Canonical Formalism

6.1.1 Generalization of the Lagrangian

In Sect. 4.3, the Lagrangian was introduced as "a function that, when substituted into the Euler-Lagrange equation, produces Newton's equation". This framework can be generalized to other "physical governing equations" by extending the definition of the Lagrangian to mean "a function that derives governing equations from the Euler-Lagrange equation". For example, the governing equation for an electrodynamic

[1] As pointed out in Sect. 1.3(c), when teaching students from fields like information science or other non-material science domains, a vague explanation such as "Spin is a magnetic moment and thus forms an inner product with the magnetic field **B**" often fails to satisfy their understanding.

R. Maezono, *Essentials for Deeper Understanding of Quantum Computing*,
https://doi.org/10.1007/978-981-96-5646-2_6

system subject to **Lorentz force**,

$$m\frac{d^2\mathbf{r}}{dt^2} = e\left(\mathbf{E} + \mathbf{v} \times \mathbf{B}\right),$$

can be derived from the Lagrangian

$$L\left(\mathbf{r}, \dot{\mathbf{r}}\right) = \frac{1}{2}m \cdot \dot{\mathbf{r}}^{\,2} - e\phi\left(\mathbf{r}\right) + e\dot{\mathbf{r}} \cdot \mathbf{A}\left(\mathbf{r}\right), \tag{6.1}$$

using the Lagrange equation of motion (4.25). Here, $\left(\phi\left(\mathbf{r}\right), \mathbf{A}\left(\mathbf{r}\right)\right)$ are the scalar potential and vector potential, which describe the electromagnetic fields $\left(\mathbf{E}\left(\mathbf{r}\right), \mathbf{B}\left(\mathbf{r}\right)\right)$ as

$$\mathbf{E}\left(\mathbf{r}, t\right) = -\nabla\phi\left(\mathbf{r}, t\right) - \frac{\partial \mathbf{A}\left(\mathbf{r}, t\right)}{\partial t}, \tag{6.2}$$

$$\mathbf{B}\left(\mathbf{r}, t\right) = \nabla \times \mathbf{A}\left(\mathbf{r}, t\right), \tag{6.3}$$

as given in (3.40). This formalism can be extended to describe the dynamics of field variables $\phi_l\left(\mathbf{r}, t\right)$ (Sect. 9.5.1). It is possible to construct a Lagrangian formulation for many examples, including elastic fields and electromagnetic fields [1, 2].

Modern Physics: Determining the Lagrangian in Reverse
In classical particle mechanics, the Lagrangian was introduced progressively through inductive reasoning, starting from Newtonian equations of motion (Sect. 4.3). The Lagrangian serves as the fundamental quantity governing the dynamics of a system and exhibits covariance under coordinate transformations as well as various symmetries. Conversely, in modern physics, which seeks to uncover unknown governing laws and equations, this reasoning is utilized in reverse. That is, by narrowing down the possible forms of the Lagrangian based on "expected requirements" such as the system's symmetry or renormalizability, a functional form is proposed as a hypothesis. The Euler-Lagrange equations derived from this functional form are then used as the "hypothesis of the governing laws" to further develop the discussion.

6.1.2 Canonical Formulation of Electromagnetic Fields

The canonical formulation based on Eq. (6.1) is utilized when discussing the electromagnetic response of material systems and serves as a foundation for subsequent discussions. Hence, let us proceed slightly further to explore its implications. The generalized momentum derived from the Lagrangian in Eq. (6.1) is expressed as:

$$\mathbf{p} = \frac{\partial L}{\partial \dot{\mathbf{r}}} = m \cdot \dot{\mathbf{r}} + e\mathbf{A}(\mathbf{r}), \tag{6.4}$$

where **p** represents the generalized momentum. Using this expression, the Hamiltonian is constructed via a Legendre transformation:

$$\begin{aligned} H(\mathbf{r}, \mathbf{p}) &= \mathbf{p} \cdot \dot{\mathbf{r}} - L(\mathbf{r}, \dot{\mathbf{r}}) \\ &= (m \cdot \dot{\mathbf{r}} + e\mathbf{A}(\mathbf{r})) \cdot \dot{\mathbf{r}} - \left(\frac{1}{2} m \cdot \dot{\mathbf{r}}^2 - e\phi(\mathbf{r}) + e\dot{\mathbf{r}} \cdot \mathbf{A}(\mathbf{r})\right) \\ &= \frac{1}{2} m \cdot \dot{\mathbf{r}}^2 + e\phi(\mathbf{r}), \end{aligned} \tag{6.5}$$

resulting in the Hamiltonian $H(\mathbf{r}, \mathbf{p})$.

It follows from the above that the motion of a charged particle ($q = -e$) in an electromagnetic field $(\phi(\mathbf{r}), \mathbf{A}(\mathbf{r}))$ is described by substituting the momentum and energy of a free classical particle as follows:

$$\mathbf{p} \to \mathbf{p} - q\mathbf{A}(\mathbf{r}), \quad E \to E - q\phi(\mathbf{r}) \tag{6.6}$$

It should be noted that the kinetic energy term in the first term of Eq. (6.1) is modified in the relativistic domain, as discussed later in Eq. (6.15), as follows:

$$\frac{1}{2} m \cdot \dot{\mathbf{r}}^2 \to -mc^2 \sqrt{1 - \frac{\dot{\mathbf{r}}^2}{c^2}}$$

Correspondingly, the first term on the far right-hand side of Eq. (6.5) is also corrected. However, it can be verified that the conclusion in Eq. (6.6) remains unchanged.

Ambiguity in the Lagrangian

At the end of Sect. 4.3.1, the concept of covariance was explained. When determining a Lagrangian in the approach where "the Euler-Lagrange equation derived from a given Lagrangian provides the governing equation", it is shown that the Lagrangian is not necessarily uniquely determined [1]. If a difference in functions, $L' = L + \Omega$, provides "the same Euler-Lagrange equations as L", then a requirement arises that "the contribution from this functional difference Ω must cancel out", leading to the condition:

(continued)

$$\frac{d}{dt}\left(\frac{\partial \Omega}{\partial \dot{q}^i}\right) - \frac{\partial \Omega}{\partial q^i} = 0$$

Using this condition as a basis for discussion, one can derive "transformations of the Lagrangian that preserve the same governing equations". This is referred to as a **gauge transformation**. When this general theory is applied to the electromagnetic interaction in Eq. (6.1), it yields the gauge transformation of electromagnetic interactions:

$$\phi(\mathbf{r}, t) \rightarrow \phi(\mathbf{r}, t) - \frac{1}{c}\frac{\partial \chi(\mathbf{r}, t)}{\partial t}, \quad \mathbf{A}(\mathbf{r}, t) \rightarrow \mathbf{A}(\mathbf{r}, t) + \nabla \chi(\mathbf{r}, t)$$

This "gauge transformation for electromagnetic fields" is familiar even in elementary electromagnetism textbooks. However, it is often introduced in a somewhat ad hoc manner with the statement, "the same equations arise even when such indeterminate elements are added". The above discussion provides a natural logical foundation behind this approach.

6.2 Overview of Relativistic Mechanics

For beginners, the most challenging aspect of special relativity is often found in the introduction of four-momentum as described in Eq. (6.14). The principle of the constancy of the speed of light is repeated in many popular science books and explained in various ways, so this concept can be digested relatively easily. Once this principle is understood, the formulation of isometric transformations, and the idea that "time and space transform into each other as components of a four-vector", are also not particularly difficult to grasp. If one accepts the concept of four-momentum, the subsequent formalism primarily involves straightforward algebraic manipulations, making it relatively simple. The reason why beginners struggle with the concept of four-momentum is discussed in the footnote around Eq. (6.14). To help beginners land softly at this point, a solution lies in starting by updating their understanding of the meaning of vectors within the context of coordinate transformation rules.

6.2.1 Vectors and Scalars

The foundation for teaching relativity lies in the perspective of "how a field transforms when the reference coordinate system changes". Let us revisit and clarify the concepts of vectors and scalars from this perspective. Consider a scenario in three-dimensional space where scalar fields (1 component) and vector fields (3 components) are defined at each point. When the coordinate axes of the reference system are rotated, scalar quantities remain unchanged, whereas vector quantities have their component representations transformed. We redefine vectors and scalars based on their transformation properties as follows [3].

The basis vector transformation of the coordinate system is given by

$$\mathbf{e}_\mathbf{j}' := \hat{R} \cdot \mathbf{e}_j = R_{kj} \cdot \mathbf{e}_k \quad \left(= \sum_k R_{kj} \cdot \mathbf{e}_k \right), \tag{6.7}$$

under which the component representation of the position vector transforms as

$$x_j' = R_{jk} \cdot x_k, \tag{6.8}$$

as shown in [(2.43) and (2.44)].[2] Thus, for the coordinate transformation given in (6.7), quantities $\{v_j\}$ that transform as $v_j' = R_{jk} \cdot v_k$ are **redefined as vectors** [3].[3]

For the coordinate transformation given in (6.7), the representation matrix $A = \{a_{ij}\}$ of the linear mapping $\mathbf{y} = \hat{A} \cdot \mathbf{x}$ is also subject to transformation. Writing this mapping in the original (unprimed) and transformed (primed) coordinate systems as

$$\mathbf{y} = \hat{A} \cdot \mathbf{x}, \quad \mathbf{y}' = \hat{A}' \cdot \mathbf{x}',$$

and substituting $\mathbf{y}' = R \cdot \mathbf{y}$, etc., yields

$$R \cdot \mathbf{y} = A' \cdot R \cdot \mathbf{x}, \quad \therefore \quad \mathbf{y} = \left(R^{-1} \cdot A' \cdot R \right) \cdot \mathbf{x} = A \cdot \mathbf{x},$$

from which we obtain the **transformation rule** $A' = R \cdot A \cdot R^{-1}$. Writing this matrix product in component form [(Eq. 2.45)], the transformation rule for the representation matrix of the projection becomes

$$a_{ij}' = R_{il} \cdot a_{lm} \cdot R_{mj}^{-1} = R_{il} \cdot a_{lm} \cdot R_{mj}^{T} = R_{il} R_{jm} \cdot a_{lm}.$$

---

[2] Note the difference in the placement of the dummy variable k in (6.7) and (6.8), appearing as $R_{j\underline{k}}$ and $R_{\underline{k}j}$, respectively, depending on whether it precedes or follows. It is crucial to correctly recognize this distinction. This difference becomes prominent in the case of oblique coordinates, where transformations are classified as covariant when the variable comes before and contravariant when it comes after [4].

[3] Under this definition, one notices that the vector cross product and scalar triple product follow transformation rules that differ from those of conventional vectors or scalars (due to an additional sign inversion along with the component transformation). These are called pseudovectors and pseudoscalars, respectively, for this reason [3].

Rewriting explicitly, for the coordinate transformation given in (6.7), the representation matrix $A = \{a_{ij}\}$ of the linear map $\mathbf{y} = \hat{A} \cdot \mathbf{x}$ transforms as

$$a'_{ij} = R_{il} R_{jm} \cdot a_{lm}.$$

Reading the arrangement of indices, each of the two legs l, m of a_{lm} corresponds to two Rs, as $R_{i\underline{l}} R_{j\underline{m}}$, where the underlined indices represent the trailing indices of R that match l, m, while the leading indices i, j represent the transformation's output in the transformed coordinate system.

Thus, quantities $a^{(N)}_{ij\cdots}$ that transform according to similar rules, where the indices are matched with the trailing indices of R and transformed to the leading indices, are defined as "N-rank **tensors**" as follows:

$$a'^{(2)}_{ij} = R_{il} R_{jm} \cdot a^{(2)}_{lm}, \quad a'^{(3)}_{ijk} = R_{il} R_{jm} R_{kn} \cdot a^{(3)}_{lmn}, \quad a'^{(4)}_{ijkl} = \cdots$$

[3, 5]. A vector is a first-rank tensor, and the representation matrix of a linear mapping is a second-rank tensor. Quantities that transform in this manner, considered to "reside in coordinate system", are referred to as vector fields and tensor fields.

In N-dimensional space, arranging arbitrary N components does not automatically constitute a vector. Instead, it is the "set of components" that transforms in correspondence with the transformation properties of the basis vectors that qualifies as a **vector**. The **inner product** formed by such vectors is

$$\begin{aligned}(\mathbf{a}' \cdot \mathbf{b}') &= a'_l b'_l = R_{lm} a_m \cdot R_{ln} b_n = R_{lm} R_{ln} \cdot a_m b_n = R^T_{ml} R_{ln} \cdot a_m b_n \\ &= \left(R^{-1} \cdot R\right)_{mn} \cdot a_m b_n = \delta_{mn} \cdot a_m b_n = a_m b_m,\end{aligned}$$

which demonstrates that the inner product of vectors constructs a scalar quantity that remains **invariant** under coordinate transformations. From the above, we can reaffirm the fundamental steps underlying relativistic mechanics:

- Identify "what components constitute a vector" under coordinate transformations—that is, determine the set of components that transform in correspondence with the basis vector transformations,
- Construct the inner product from the identified vectors to derive invariant quantities,

With these principles firmly recognized, we proceed to the next discussion.

6.2.2 Relativistic Dynamics

When signals or information propagate at the speed c, the distance traveled by the signal during a time interval dt in a coordinate system (x, y, z) is expressed as:

$$dx^2 + dy^2 + dz^2 = c^2 \cdot dt^2.$$

Now consider capturing this event in another coordinate system $\left(x', y', z'\right)$ that is moving uniformly at a finite relative velocity.

Under conventional reasoning, the relative velocity changes the observed propagation speed of c to c', such that:

$$dx^2 + dy^2 + dz^2 = c^2 \cdot dt^2, \quad dx'^2 + dy'^2 + dz'^2 = c'^2 \cdot dt^2.$$

Implicit in this reasoning is the assumption that "the time t flows identically and invariantly across different coordinate systems". However, **principle of the invariance of the speed of light** asserts that the maximum propagation speed of signals or information, c, remains identical and invariant across all coordinate systems (Sect. 9.5.3). Consequently, the above equations should instead be written as:

$$dx^2 + dy^2 + dz^2 = c^2 \cdot dt^2, \quad dx'^2 + dy'^2 + dz'^2 = c^2 \cdot dt'^2. \tag{6.9}$$

The above equations imply that it is not the time t that is invariant, but rather the speed of light c. Instead, time transforms similarly to spatial coordinates as $t \to t'$. This signifies that we adopt a perspective where **spacetime** (x, y, z, t), which incorporates time into the spatial coordinates, undergoes transformations in a four-dimensional space. This framework allows us to conceptualize phenomena under the view that spacetime coordinates transform together. The seemingly peculiar recognition of the invariance of the speed of light, and the understanding that the maximum propagation speed c is indeed the speed of light, are briefly elaborated in Sect. 9.5.3.

From Eq. (6.9), we can derive the velocity addition formula (Sect. 9.5.2). When a velocity v observed in the stationary frame is viewed from a frame moving with a relative velocity V, the velocity v' observed in the moving frame is related to v by

$$v = \frac{v' + V}{1 + v'\left(V/c^2\right)}. \tag{6.10}$$

In the limit $V \ll c$ (non-relativistic limit), the denominator approaches 1, reducing to the familiar velocity addition formula $v' = v + V$. Additionally, when $v = c$, we find that $v' = c$, as expected.

Equation (6.9) can be rewritten as

$$0 = c^2 \cdot dt^2 - \left(dx^2 + dy^2 + dz^2\right) = c^2 \cdot dt'^2 - \left(dx'^2 + dy'^2 + dz'^2\right), \tag{6.11}$$

and thus we formulate that the spacetime transformation $(x, y, z, t) \rightarrow (x', y', z', t')$ is governed to preserve this equality. The preserved norm is written as

$$\begin{aligned} dl^2 &= c^2 \cdot dt^2 - \left(dx^2 + dy^2 + dz^2\right) \\ &= (dx, dy, dz, c \cdot dt) \begin{pmatrix} -1 & 0 & 0 & 0 \\ 0 & -1 & 0 & 0 \\ 0 & 0 & -1 & 0 \\ 0 & 0 & 0 & 1 \end{pmatrix} \begin{pmatrix} dx \\ dy \\ dz \\ c \cdot dt \end{pmatrix} \\ &= dx^\mu \cdot g_{\mu\nu} \cdot dx^\nu, \end{aligned} \tag{6.12}$$

in the **isometric transformation**. The four-component vector dx^ν corresponds to the spacetime coordinate components in four-dimensional spacetime, and since dl^2 is invariant, the above expression defines the "inner product operation that forms a scalar invariant from a vector". For ordinary three-dimensional vectors, we have

$$\begin{aligned} \left[dl^{(3)}\right]^2 &= dx^2 + dy^2 + dz^2 \\ &= (dx, dy, dz) \begin{pmatrix} 1 & 0 & 0 \\ 0 & 1 & 0 \\ 0 & 0 & 1 \end{pmatrix} \begin{pmatrix} dx \\ dy \\ dz \end{pmatrix} \\ &= dx^\mu \cdot \delta_{\mu\nu} \cdot dx^\nu, \end{aligned}$$

but in four-dimensional spacetime, the inner product is defined by replacing the 3×3 identity matrix $\delta_{\mu\nu}$ with the 4×4 metric tensor $g_{\mu\nu}$. From the above isometric transformation condition, we obtain, for example, the spacetime transformations called **Lorentz transformations**:

$$x = \frac{x' + vt'}{\sqrt{1 - (v/c)^2}}, \quad t = \frac{x' \cdot (v/c^2) + t'}{\sqrt{1 - (v/c)^2}}, \tag{6.13}$$

as derived in [Sect. 9.5.2].

The invariant quantity of the isometric transformation, given by Eq. (6.12), can be expressed as

$$dl^2 = c^2 \cdot dt^2 \left(1 - \frac{1}{c^2 \cdot dt^2}\left(dx^2 + dy^2 + dz^2\right)\right),$$

$$\therefore \quad dl = c \cdot dt \sqrt{1 - \frac{v^2}{c^2}} =: c \cdot ds,$$

where c and dl are invariant under spacetime transformations. From this, we see that

$$ds = dt\sqrt{1 - \frac{v^2}{c^2}}$$

is also an invariant quantity under spacetime transformations. The quantity ds is called the **proper time**, and in the non-relativistic limit $v \ll c$, it reduces to $ds \to dt$, reverting to the familiar infinitesimal time. This implies that the non-relativistic assumption of time invariance, where dt was considered invariant, is corrected. Instead, we recognize that ds is invariant, and in the non-relativistic limit, $ds \to dt$, giving the appearance that dt is invariant.

In non-relativistic cases, dt was considered an invariant quantity under coordinate transformations. Thus, dividing the vector $\mathbf{r}$ by this invariant, $d\mathbf{r}/dt$, produced a quantity that could be **treated as a vector**, transforming according to the same rules as $\mathbf{r}$. However, in the relativistic case, constructing a four-component quantity dx^μ/dt from the vector $dx^\mu = (dx, dy, dz, c \cdot dt)^T$ becomes problematic, as dt is no longer invariant and changes depending on the reference frame. Referring to Sect. 6.2.1, such a quantity is "**merely a collection of numbers, not a vector**". To resolve this, we replace dt with ds and define

$$\frac{dx^\mu}{ds} = \frac{1}{dt\sqrt{1 - \frac{v^2}{c^2}}}\begin{pmatrix} dx \\ dy \\ dz \\ c \cdot dt \end{pmatrix} = \frac{1}{\sqrt{1 - \frac{v^2}{c^2}}}\begin{pmatrix} v_x \\ v_y \\ v_z \\ c \end{pmatrix},$$

which becomes a vector quantity.[4] We then define the **four-momentum vector** by multiplying this four-velocity vector by the rest mass m_0, as follows:

$$p^\mu := m_0\frac{dx^\mu}{ds} = \frac{1}{\sqrt{1 - \frac{v^2}{c^2}}}\begin{pmatrix} m_0\mathbf{v} \\ m_0 c \end{pmatrix}. \tag{6.14}$$

[4] To facilitate a smooth transition through this section, Sect. 6.2.1 has been included as preparatory material. For beginners, the notion of "vectors as objects governed by specific transformation rules under coordinate changes" is often not well understood. Instead, many hold the misconception that a "vector = (a collection of numbers)", without having updated their conceptual framework. As a result, when introduced to the concept of dividing by proper time ds to define quantities, they may question, "Why is it necessary to divide by ds? Can't we just divide by dt?". This sudden jump in abstraction can feel abrupt, preventing subsequent discussions from being fully comprehended.

The spatial component

$$\mathbf{p} = \frac{m_0 \mathbf{v}}{\sqrt{1 - \frac{v^2}{c^2}}}$$

indeed reduces to the conventional momentum in the non-relativistic limit.

Next, the relativistic extension of the Lagrangian is introduced in correspondence with the "spatial momentum extended relativistically". Recalling the relationship between momentum and Lagrangian (4.23), we see that the Lagrangian should be extended relativistically so that the "spatial momentum extended relativistically" is given by

$$\mathbf{p} = \frac{\partial L}{\partial \mathbf{v}}.$$

Such a relativistic Lagrangian is given by

$$L = T = -m_0 c^2 \sqrt{1 - \frac{v^2}{c^2}}. \tag{6.15}$$

The Hamiltonian, in correspondence, is expressed as

$$H = -L + \mathbf{p} \cdot \mathbf{v} = m_0 c^2 \sqrt{1 - \frac{v^2}{c^2}} + \frac{m_0 v^2}{\sqrt{1 - \frac{v^2}{c^2}}} = \frac{m_0 c^2}{\sqrt{1 - \frac{v^2}{c^2}}} = E.$$

By comparing this expression for relativistic energy with the fourth component of the four-momentum vector $p^{(\mu=4)}$ in Eq. (6.14), we find

$$p^{(\mu=4)} = \frac{m_0 c}{\sqrt{1 - \frac{v^2}{c^2}}} = \frac{E}{c}. \tag{6.16}$$

Thus, the four-momentum vector can be expressed as

$$p^{\mu} = m_0 \frac{dx^{\mu}}{ds} = \begin{pmatrix} \mathbf{p} \\ E/c \end{pmatrix}.$$

This four-component quantity "transforms as a vector", which is the primary conclusion derived so far.

Thus, the inner product formed from the four-momentum vector, as in the operation of (6.12), is a "scalar quantity invariant under coordinate transformations". This invariant quantity is given by

$$C = \left|p^{\mu}\right|^2 = p^{\mu} g_{\mu\nu} p^{\nu} = (E/c)^2 - p^2 = \left(E'/c\right)^2 - p'^2.$$

By choosing the primed frame to be the rest frame with $\mathbf{p} = 0$, we find

$$(E/c)^2 - p^2 = \left(E_{(\mathbf{p}=0)}/c\right)^2, \quad \therefore \quad E^2 - c^2 p^2 = E^2_{(\mathbf{p}=0)}.$$

From (6.16),

$$E_{(\mathbf{p}=0)} = m_0 c^2,$$

we obtain

$$E^2 - c^2 p^2 = m_0^2 c^4, \quad \therefore \quad E^2 = c^2 p^2 + m_0^2 c^4. \tag{6.17}$$

The second term, $E_{(\mathbf{p}=0)} = m_0 c^2$, is referred to as the rest energy.

The above discussion outlines the main points of relativistic mechanics. Let us briefly review the flow of the arguments to consolidate our understanding before transitioning to the next section on "relativistic quantum mechanics." We have aimed to derive (6.17) as the final goal of this discussion. To achieve this, we started by postulating the "principle of invariance of the speed of light". This postulate was then formulated as the "isometric transformations that preserve the inner product of four-vectors". We defined four-velocity and four-momentum as vectors that transform under these isometric transformations by normalizing spacetime four-vectors with the invariant scalar, proper time. Using the spatial component of the four-momentum as a starting point, we derived relativistic extensions of the Lagrangian and Hamiltonian formulations. Building upon this foundation, we established the relationship between four-momentum and energy. Finally, through the transformation properties of four-momentum as a vector, we arrived at the energy relation given by (6.17).

6.3 Relativistic Quantum Mechanics

Historically, it became known through experimental observations that the spectral lines of atoms split when subjected to a magnetic field (the Zeeman effect). This led to the establishment of the concept that electrons possess an intrinsic degree of freedom, namely spin, which can align either upward or downward in response to an external magnetic field. This "two-component internal degree of freedom" can be derived from relativistic quantum mechanics, as discussed in this section.

6.3.1 Dirac Equation

As derived in the previous section, the fundamental quantities of classical particle mechanics, E and $\mathbf{p}$, must satisfy the relativistic requirement

$$E^2 = m^2c^4 + p^2c^2, \tag{6.18}$$

as was established earlier.[5] When attempting to construct relativistic quantum mechanics, one idea is to rewrite Eq. (6.18) into a form like $E = \cdots$ and, using the correspondence principle in Eq. (5.9), arrive at

$$-\frac{\hbar}{i}\frac{\partial}{\partial t}\psi(\mathbf{r}, t) = \cdots, \tag{6.19}$$

as a "description of the time evolution of the wave function".

To express Eq. (6.18) in the form of "$E = \cdots$", if we naively take the square root of both sides, the right-hand side of the equation will contain p^2 within the square root. However, as seen in the Lorentz transformation Eq. (6.13), relativity requires spacetime to mix linearly and equivalently. Therefore, it is reasonable to expect that "$\mathbf{p} \sim \nabla$" should appear linearly on the right-hand side of Eq. (6.19). The above-mentioned "square root expression" does not align with this expectation. Thus, the discussion shifts toward seeking a **transformation that avoids taking the square root**.

From the dimensional relation of Eq. (6.18),

$$E \sim c \cdot p + mc^2, \tag{6.20}$$

we consider whether squaring this expression can satisfy the relationship in Eq. (6.18). For usual "single-element numbers", cross terms appear, making this infeasible. However, through concepts from binaries, quaternions, or other hypercomplex numbers [6], we can construct a framework satisfying Eq. (6.18) by imposing algebraic relations that prevent cross terms. Specifically, let us rewrite Eq. (6.20) as

$$E = \left[+c \cdot \underline{\boldsymbol{\alpha}} \cdot \mathbf{p} + \underline{\beta} \cdot mc^2\right], \tag{6.21}$$

where coefficients $\underline{\boldsymbol{\alpha}}$ and $\underline{\beta}$ from hypercomplex algebra are appended to each term. These coefficients satisfy the conditions

$$\underline{\beta} \cdot \underline{\beta} = 1, \quad \left(\underline{\alpha_l} \cdot \underline{\beta} + \underline{\beta} \cdot \underline{\alpha_l}\right) = 0, \quad \left(\underline{\alpha_k} \cdot \underline{\alpha_l} + \underline{\alpha_l} \cdot \underline{\alpha_k}\right) = 2\delta_{kl}. \tag{6.22}$$

[5] In this section, there is no risk of confusion in writing the rest mass m_0 as m, and thus it is noted accordingly.

If these algebraic relations hold, squaring Eq. (6.21) yields the relativistic requirement in Eq. (6.18).

The conditions in Eq. (6.22) can be satisfied using 4×4 complex matrices, specifically:

$$\underline{\beta} = \begin{pmatrix} \underline{I} & 0 \\ 0 & -\underline{I} \end{pmatrix}, \quad \underline{\alpha}_j = \begin{pmatrix} 0 & \underline{\sigma_j} \\ \underline{\sigma_j} & 0 \end{pmatrix},$$

as shown in [7]. Here, $\underline{I}$ is the identity matrix, and

$$\sigma_x = \begin{pmatrix} 0 & 1 \\ 1 & 0 \end{pmatrix}, \quad \sigma_y = \begin{pmatrix} 0 & -i \\ i & 0 \end{pmatrix}, \quad \sigma_z = \begin{pmatrix} 1 & 0 \\ 0 & -1 \end{pmatrix} \tag{6.23}$$

are known as **Pauli matrices**. These Pauli matrices hold specific significance, as will be discussed in Sect. 7.1. At this stage, however, note that their introduction is solely for **mathematical convenience**.[6]

As a result, we have successfully reformulated the relativistic requirement in Eq. (6.18) into a linear form $E = \cdots$, without using square roots. However,

$$\underline{E} = \left[+c \cdot \underline{\boldsymbol{\alpha}} \cdot \mathbf{p} + \underline{\beta} \cdot mc^2\right] = \left[\begin{pmatrix} mc^2\underline{1} & c\underline{\sigma} \cdot \mathbf{p} \\ c\underline{\sigma} \cdot \mathbf{p} & -mc^2\underline{1} \end{pmatrix}\right]$$

takes the form of a 4×4 complex matrix.[7] Applying the correspondence principle in Eq. (5.9) to this expression replaces $\mathbf{p}$ with the operator $\hat{\mathbf{p}}$, yielding

$$\left[\begin{pmatrix} mc^2\underline{1} & c\underline{\sigma} \cdot \hat{\mathbf{p}} \\ c\underline{\sigma} \cdot \hat{\mathbf{p}} & -mc^2\underline{1} \end{pmatrix}\right] |\psi^{(4)}\rangle = E \cdot |\psi^{(4)}\rangle, \tag{6.24}$$

which represents a governing equation acting on the wavefunction $|\psi\rangle$ (known as the **Dirac equation**). Here, $|\psi^{(4)}\rangle$ is given by

$$|\psi^{(4)}\rangle = \begin{pmatrix} |\psi_L^{(2)}\rangle \\ |\psi_S^{(2)}\rangle \end{pmatrix},$$

a four-component vector. For later convenience, this is expressed in terms of the two-component vectors $|\psi_{L,S}^{(2)}\rangle$.

[6] They were introduced as an algebra that avoids cross terms.

[7] While written in a 2×2 block form, each element itself is a 2×2 matrix, resulting in a 4×4 matrix.

6.3.2 Coupling with Magnetic Fields

In the canonical formulation of electromagnetic fields (Hamiltonian-based description) the effects of the electromagnetic field can be described through the substitution

$$\mathbf{p} \rightarrow \mathbf{p} - q\mathbf{A}(\mathbf{r}), \quad E \rightarrow E - q\phi(\mathbf{r}) \tag{6.25}$$

as explained in Eq. (6.6). Applying this substitution to Eq. (6.24) and writing it explicitly yields a pair of coupled equations for $|\psi^{(2)}_{L,S}\rangle$:

$$\begin{aligned} c\underline{\sigma}\cdot\left(\hat{\mathbf{p}} - q\mathbf{A}\right)|\psi^{(2)}_S\rangle &= \left(E - q\phi - mc^2\right)|\psi^{(2)}_L\rangle \\ c\underline{\sigma}\cdot\left(\hat{\mathbf{p}} - q\mathbf{A}\right)|\psi^{(2)}_L\rangle &= \left(E - q\phi + mc^2\right)|\psi^{(2)}_S\rangle. \end{aligned} \tag{6.26}$$

These equations describe the behavior of electrons in the relativistic regime.[8] This formula is known as the **Dirac equation**.

The rest energy $E_0 = mc^2$ included in the relativistic energy can be separated as

$$E = mc^2 + E_{\mathrm{NR}} \tag{6.27}$$

where E_{NR} represents "the energy considered in non-relativistic theory".[9] In this context, the non-relativistic limit corresponds to the condition

$$mc^2 \gg E_{\mathrm{NR}}, \quad mc^2 \gg q\phi, \quad E \approx mc^2, \tag{6.28}$$

where mc^2 overwhelmingly exceeds other energy scales. If the Dirac equation [Eq. (6.26)] is asymptotically evaluated in this limit, it is expected to reduce to the non-relativistic Schrodinger equation. Indeed, as shown below, a form close to the Schrodinger equation is obtained, but it comes with "additional terms". These additional terms describe "the coupling between spin and magnetic fields", as explained later.

By substituting Eq. (6.27) into the E term on the right-hand side of Eq. (6.26), and taking the limit as described in Eq. (6.28), we obtain

$$\begin{aligned} c\underline{\sigma}\cdot\left(\hat{\mathbf{p}} - q\mathbf{A}\right)|\psi^{(2)}_S\rangle &= \left(E_{\mathrm{NR}} - q\phi\right)|\psi^{(2)}_L\rangle \\ c\underline{\sigma}\cdot\left(\hat{\mathbf{p}} - q\mathbf{A}\right)|\psi^{(2)}_L\rangle &\approx 2mc^2|\psi^{(2)}_S\rangle. \end{aligned}$$

[8] The formulation satisfying the relativistic requirement in Eq. (6.18) is not unique [7–9]. The "4-dimensional algebraic formulation" discussed here is specifically a theory for describing electrons. Other formulations describe other particles.

[9] The subscript 'NR' stands for Non-relativistic.

Comparing the right-hand sides of the first and second equations, it is evident that the second equation's right-hand side is overwhelmingly large, setting the scale for the magnitude of $|\psi_{L,S}^{(2)}\rangle$. In response to this disparity, the subscripts on ψ, (L = Large) and (S = Small), are defined to reflect their relative magnitudes. Since the second equation does not include the unknown E_{NR}, it can be solved as

$$|\psi_S^{(2)}\rangle \approx \frac{1}{2mc}\underline{\sigma} \cdot (\hat{\mathbf{p}} - q\mathbf{A})\,|\psi_L^{(2)}\rangle.$$

This solution for $|\psi_S^{(2)}\rangle$ can then be substituted into the first equation of Eq. (6.26), eliminating the subdominant $|\psi_S^{(2)}\rangle$. In this way, we obtain an effective equation for the dominant $|\psi_L^{(2)}\rangle$:

$$\left[\frac{1}{2m}\underline{\sigma} \cdot (\hat{\mathbf{p}} - q\mathbf{A})\,\underline{\sigma} \cdot (\hat{\mathbf{p}} - q\mathbf{A}) + q\phi \cdot \underline{1}\right] |\psi_L^{(2)}\rangle = E_{\mathrm{NR}} \cdot |\psi_L^{(2)}\rangle. \tag{6.29}$$

This yields a two-component equation describing the effective dynamics.

For the Pauli matrices in Eq. (6.23), the property [Eq. (9.32) in Sect. 9.5.4]

$$\sigma_j \sigma_k = \delta_{jk} + i \cdot \varepsilon_{jkl} \cdot \sigma_l \tag{6.30}$$

can be applied to rewrite the first term in Eq. (6.29) as

$$\sigma \cdot (\mathbf{p} - q\mathbf{A})\,\sigma \cdot (\mathbf{p} - q\mathbf{A}) = (\mathbf{p} - q\mathbf{A})^2 \cdot \underline{1} - q\hbar\,\sigma \cdot (\nabla \times \mathbf{A})\,, \tag{6.31}$$

derived as Eq. (9.34) in Sect. 9.5.4]. Since $\mathbf{B} = \nabla \times \mathbf{A}$, Eq. (6.29) ultimately reduces to

$$\left[\frac{1}{2m} \cdot (\mathbf{p} - q\mathbf{A})^2 \cdot \underline{1} - \frac{q\hbar}{2m}\underline{\sigma} \cdot \mathbf{B} + q\phi \cdot \underline{1}\right] |\psi_L^{(2)}\rangle = E_{\mathrm{NR}} \cdot |\psi_L^{(2)}\rangle. \tag{6.32}$$

Equation (6.32) fundamentally shares the same form as the Schrodinger equation in Eq. (5.10) with the substitution for the electromagnetic field in Eq. (6.25):

$$\left[\frac{1}{2m} \cdot (\mathbf{p} - q\mathbf{A})^2\right] |\psi\rangle = (E_{\mathrm{NR}} - q\phi) \cdot |\psi\rangle,$$

but it is further expressed in a two-component form[10] with an additional term proportional to $\underline{\sigma} \cdot \mathbf{B}$. If there is no magnetic field, i.e., $\mathbf{B} = 0$, the 2×2 Hamiltonian matrix becomes diagonal, and the two components of the wave function $|\psi_L^{(2)}\rangle$ do not mix, reducing to the non-relativistic Schrodinger equation for each component individually.

[10] The state $|\psi_L^{(2)}\rangle$ was defined as a two-component vector.

If the magnetic field **B** is directed along the z-axis, the diagonal matrix σ_z from Eq. (6.23) appears as an additional term proportional to the magnetic field strength $|B|$. In this case, the 2×2 Hamiltonian matrix becomes diagonal, and the two components of the wave function $|\psi_L^{(2)}\rangle$ remain independent, without mixing. Each component receives a contribution proportional to $|B|$, but with opposite signs. As a result, the energy levels corresponding to each component shift in opposite directions. This explains the Zeeman splitting, which was already known experimentally [10].

6.3.3 State Control by External Magnetic Fields

When **B** is oriented along the x or y direction, the Pauli matrices $\sigma_{x,y}$, which contain off-diagonal components, are added to the Hamiltonian. As a result, the 2×2 Hamiltonian matrix introduces off-diagonal elements, causing the two components of the wave function $|\psi_L^{(2)}\rangle$ to mix with each other.

The commonly provided explanation here is as follows. This explanation is highly unintuitive for students who are proficient in mathematics but beginners in physics. It states, "Since the additional term appears in the form of the inner product $-\underline{\sigma} \cdot \mathbf{B}$, $\underline{\sigma}$ tends to align with **B**".[11] When **B** points in the x direction, $\underline{\sigma}$ becomes polarized along the x axis. Thus, based on the Bloch sphere illustration [Fig. 1.1a] and the explanation of Eq. (1.2), it is said that this results in the superposition state $|\psi_{\pi/2,0}^{(2)}\rangle \sim (|0\rangle + |1\rangle)$. For students with prior experience in quantum physics, this explanation might elicit an "Ah, I remember that" response, allowing them to proceed through the material by relying on their familiarity. However, upon closer inspection, the triplet of matrices $\underline{\sigma} = \{\sigma_x, \sigma_y, \sigma_z\}$, introduced in Eq. (6.23), were only defined as a mathematical convenience to eliminate cross terms in transforming $E^2 = \cdots$ into $E = \cdots$. Their relationship to the "directional vector $\boldsymbol{\sigma}$" depicted in Fig. 1.1a and $|\psi_{\theta,\phi}^{(2)}\rangle$ has not been explained anywhere thus far. Hence, for students from other fields, the above explanation may lack sufficient logical coherence to advance comfortably.[12]

[11] This additional term contributes to the Hamiltonian, that is, the energy of the system. The inner product $\mathbf{A} \cdot \mathbf{B}$ is maximized when **A** and **B** point in the same direction. Therefore, the negative additional term $-\underline{\sigma} \cdot \mathbf{B}$ minimizes the system's energy when **B** and $\underline{\sigma}$ are aligned.

[12] It is often believed that the primary challenge for (reasonably thoughtful) beginners from other fields struggling with mastering quantum computing lies in the "superposition states", famously labeled as "incomprehensible even by Feynman". However, it seems more likely that accepting such phenomena as "just the way things are" is not particularly difficult for them. Rather, the issue appears to lie in the fact that most texts omit detailed explanations of the logical progression at this stage, which impedes their understanding, a perspective informed by lecture experiences. The "content that should not be omitted and must be explained carefully" is, as discussed in the next section, more a matter of "classical mathematics" than "quantum physics". For students in information sciences with a solid grasp of logic, this is a topic they can readily comprehend.

A slightly more satisfactory and detailed explanation might proceed as follows. Reflecting on the two-component Schrodinger equation (6.32), it is written as a 2×2 matrix eigenvalue equation:

$$\hat{H} = \left[C_1 \cdot \underline{1} - C_2 \cdot \underline{\sigma} \cdot \mathbf{B}\right], \quad \hat{H}|\psi_L^{(2)}\rangle = E_{\mathrm{NR}} \cdot |\psi_L^{(2)}\rangle. \tag{6.33}$$

The two-component eigenvector that satisfies this equation determines the "state realized in the system". Since the C_1 term is diagonal, the eigenvectors of $\hat{H}$ are those of the C_2 term, which includes the off-diagonal elements (see Sect. 9.5.5). For example, when a transverse magnetic field $\mathbf{B} \propto \mathbf{e_x}$ is applied, the C_2 term becomes $C_2 B \cdot \sigma_x$, and the state realized in the system corresponds to the eigenstates of σ_x.

As shown in Sect. 9.5.6, the eigenstates corresponding to σ_x, σ_y, and σ_z, with eigenvalues denoted by λ, are given as follows:

$$|\psi\rangle_{\lambda=1}^{(z)} = |1\rangle, \quad |\psi\rangle_{\lambda=-1}^{(z)} = |0\rangle$$

$$|\psi\rangle_{\lambda=1}^{(x)} = \frac{1}{\sqrt{2}}(|0\rangle + |1\rangle), \quad |\psi\rangle_{\lambda=-1}^{(x)} = \frac{1}{\sqrt{2}}(|0\rangle - |1\rangle)$$

$$|\psi\rangle_{\lambda=1}^{(y)} = \frac{1}{\sqrt{2}}(|0\rangle + i \cdot |1\rangle), \quad |\psi\rangle_{\lambda=-1}^{(y)} = \frac{1}{\sqrt{2}}(|0\rangle - i \cdot |1\rangle).$$

Since the Hamiltonian has a negative sign in $H \sim -\boldsymbol{\sigma} \cdot \mathbf{B}$, the eigenstates corresponding to the higher eigenvalue $\lambda = +1$ represent the stable states with the lowest energy. These states are given as follows:

$$|\psi\rangle_{\lambda=1}^{(z)} = |1\rangle, \quad |\psi\rangle_{\lambda=1}^{(x)} = \frac{1}{\sqrt{2}}(|0\rangle + |1\rangle), \quad |\psi\rangle_{\lambda=1}^{(y)} = \frac{1}{\sqrt{2}}(|0\rangle + i \cdot |1\rangle).$$

Focusing specifically on $|\psi\rangle_{\lambda=1}^{(x)}$, it is evident that, when a transverse magnetic field $\mathbf{B} \propto \mathbf{e_x}$ is applied, a superposition state known as the Hadamard state,

$$|\psi\rangle_{\lambda=1}^{(x)} = \frac{1}{\sqrt{2}}(|0\rangle + |1\rangle),$$

is realized.

In summary, the additional term $-\underline{\sigma} \cdot \mathbf{B}$ alters the off-diagonal components of the Hamiltonian matrix through the external magnetic field $\mathbf{B}$ and thereby controls the lowest-energy eigenstate $|\psi_L^{(2)}\rangle$. Thus, the polarization direction of $\underline{\sigma}$ can be identified with the realized lowest-energy eigenstate $|\psi_L^{(2)}\rangle$. This explains the concise statement: "When a transverse magnetic field $\mathbf{B}$ is applied in the x-direction, $\underline{\sigma}$ becomes polarized in the x-direction, and the state $|\psi_{\pi/2,0}^{(2)}\rangle \sim (|0\rangle + |1\rangle)$ is realized".

In general, when $\underline{\boldsymbol{\sigma}}$ is polarized in the direction

$$\underline{\boldsymbol{\sigma}} \propto \begin{pmatrix} n_x \\ n_y \\ n_z \end{pmatrix} = \begin{pmatrix} \sin\theta\cos\phi \\ \sin\theta\sin\phi \\ \cos\theta \end{pmatrix}$$

by the external magnetic field, the C_2 term becomes

$$\begin{aligned} -\mathbf{B}\cdot\boldsymbol{\sigma} &= -B\left(\sin\theta\cos\phi\cdot\sigma_x + \sin\theta\sin\phi\cdot\sigma_y + \cos\theta\cdot\sigma_z\right) \\ &\propto \begin{pmatrix} \cos\theta & \sin\theta\cdot e^{-i\phi} \\ \sin\theta\cdot e^{i\phi} & -\cos\theta \end{pmatrix}. \end{aligned}$$

The eigenvector of this is given by

$$|\psi^{(2)}_{(\theta,\phi)}\rangle = \begin{pmatrix} \cos(\theta/2) \\ e^{i\phi}\sin(\theta/2) \end{pmatrix}.$$

This demonstrates that the quantum state $|\psi^{(2)}_{(\theta,\phi)}\rangle$ can be identified with the direction of the spin $\boldsymbol{\sigma}$ as depicted on the Bloch sphere (Sect. 9.5.7).

Behind the approach of identifying the 2-dimensional variable $|\psi^{(2)}_{(\theta,\phi)}\rangle$ with the 3-dimensional vector $\boldsymbol{\sigma}$, lies the mathematical framework describing how a field represented by an N-dimensional basis transforms under coordinate rotation in 3-dimensional space. The case of $N = 2$ corresponds to the discussion presented here. In the next chapter, we will explain this mathematical framework, aiming to solidify the understanding of Pauli matrices.

References

1. Goldstein H, Poole CP, Safko JL (2013) Classical mechanics. Pearson, Harlow. ISBN: 978-1292026558
2. Landau LD, Lifshitz EM (1980) The classical theory of fields. Pergamon, New York. ISBN: 978-0080250724
3. Riley KF, Hobson MP, Bence SJ (2006) Mathematical methods for physics and engineering, 3rd edn. Cambridge University Press, Cambridge. ISBN: 978-0521679718
4. Fleisch D (2011) A student's guide to vectors and tensors. Cambridge University Press, Cambridge. ISBN: 978-0521171908
5. Panofsky WKH, Phillips M (2005) Classical electricity and magnetism, 2nd edn. Dover, New York. ISBN: 978-0486439242
6. Needham T (1999) Visual complex analysis. Clarendon Press, Oxford. ISBN: 978-0198534464
7. Strange P (1988) Relativistic quantum mechanics. Cambridge University Press, Cambridge. ISBN: 978-0521562713
8. Tomonaga S (1998) The story of spin. University of Chicago Press, Chicago. ISBN: 978-0226807942
9. Weinberg S (2015) Lectures on quantum mechanics, 2nd edn. Cambridge University Press, Cambridge. ISBN: 978-1107111660
10. Mansfield MM, O'Sullivan C (2020) Understanding physics, 3rd edn. Wiley, New York. ISBN: 978-1119519508

Chapter 7
Field Transformations and Spin

As stated in Chap. 1, this chapter is structured with the goal of ensuring that "readers gain confidence in their understanding of the Bloch sphere". Many books attempt to explain the essence of spin based on spinor fields, but these typically assume a higher level of prior knowledge, making them rather difficult and intimidating for beginners. In this book, I have made every effort to avoid abrupt definitions and introductions, instead building explanations using only the material covered so far. We begin with the discussion of "how the component representation of a field changes when a rotational transformation is applied to a three-dimensional coordinate system", which leads to the introduction of the concept of spin in fields (Sect. 7.1). Next, we analyze how the "two-component wavefunction in the non-relativistic limit of relativistic electron theory", introduced in Sect. 6.3, transforms under rotations of the three-dimensional coordinate system. This naturally leads to the introduction of spinor fields (Sect. 7.2). We then carefully explain what it means to "identify the spin of a field with the three-dimensional orientation illustrated on the Bloch sphere". Finally, through a clear understanding of transverse spin states, we reaffirm the meaning of "controlling spin orientation via an applied magnetic field" (Sect. 7.3).

7.1 Field Transformations and Spin

In the previous chapter, we discussed the coupling term $H \sim -\mathbf{B}\cdot\boldsymbol{\sigma}$, which appears in the non-relativistic asymptotic form of the Dirac equation. This led to the explanation that "the 3-dimensional orientation of $\boldsymbol{\sigma}$ controls the distribution of the components of the 2-component wave function" as expressed by

$$|\psi_S^{(2)}\rangle = c_0\cdot|0\rangle + c_1\cdot|1\rangle = \begin{pmatrix} c_0 \\ c_1 \end{pmatrix}. \tag{7.1}$$

R. Maezono, *Essentials for Deeper Understanding of Quantum Computing*,
https://doi.org/10.1007/978-981-96-5646-2_7

From this, we concluded that "the 3-dimensional orientation of $\boldsymbol{\sigma}$ and the state of $|\psi_S^{(2)}\rangle$ can be identified". In this chapter, we go beyond the explanation of "identification through control" to provide a deeper understanding of what it means for the 2-component field $|\psi_S^{(2)}\rangle$ to "be oriented in the direction of $\mathbf{B}$ in 3-dimensional space".

Equation (7.1) appears to resemble a "two-dimensional vector," but it is not a position vector in a two-dimensional space. Instead, it represents a "two-component field" that exists in three-dimensional space,[1] and must be treated following the precise discussion of "transformation properties under coordinate changes" outlined in Sect. 6.2.1. When the coordinate system is rotated around the **B**-axis in three-dimensional space, this "two-component field" adjusts by exchanging its components accordingly. In Sect. 2.3, we introduced the concept of representation matrices for operations such as rotations. For "coordinate rotations in three-dimensional space", we can define a "representation matrix for two-component quantities" that describes how the spinor field interchanges its two components. We will see that this representation matrix corresponds to the **Pauli matrices** introduced in the previous chapter.

7.1.1 Operators for Infinitesimal Rotations of Fields

Let us consider a multi-component field $\psi(\mathbf{r})$ defined on 3-dimensional space.[2] Under a coordinate transformation $\mathbf{r}' = \hat{R} \cdot \mathbf{r}$, the transformation law for the field $\psi' = \hat{R} \cdot \psi$ must ensure that the field retains its meaning. This requires the transformation to satisfy

$$\psi'\left(\mathbf{r}'\right) = \psi(\mathbf{r}), \quad \text{i.e.,} \quad \left[\hat{R}\psi\right](\mathbf{r}) = \psi\left(\hat{R}^{-1}\mathbf{r}\right), \tag{7.2}$$

so that the "landscape of the field remains identical around the transformed coordinates".

This can be applied to the case of a "small-angle rotation ε around the z-axis, $\hat{R}_z(\varepsilon)$". Following Eq. (7.2), we write:

$$\left[\hat{R}_z(\varepsilon) \cdot \psi\right](\mathbf{r}) = \psi\left(\hat{R}_z^{-1}(\varepsilon) \cdot \mathbf{r}\right) = \psi(\mathbf{r}_0) = \psi(x_0, y_0, z_0), \tag{7.3}$$

and noting the positional relationship in Fig. 7.1, the relationship between the new position $\mathbf{r} = (\rho, \varphi)$ and the original position $\hat{R}_z^{-1}(\varepsilon) \cdot (\mathbf{r})$ is given by

$$x_0 = \rho \cdot \cos(\varphi - \varepsilon), \quad y_0 = \rho \cdot \sin(\varphi - \varepsilon), \quad z_0 = z.$$

[1] This refers to a field expressed as an expansion in two basis function systems.

[2] Examples include p-orbitals as 3-component fields or d-orbitals as 5-component fields. The discussion in this subsection ultimately aims to determine how the transformation laws for the 2-component field $|\psi_L^{(2)}\rangle$ in Eq. (6.32) behave.

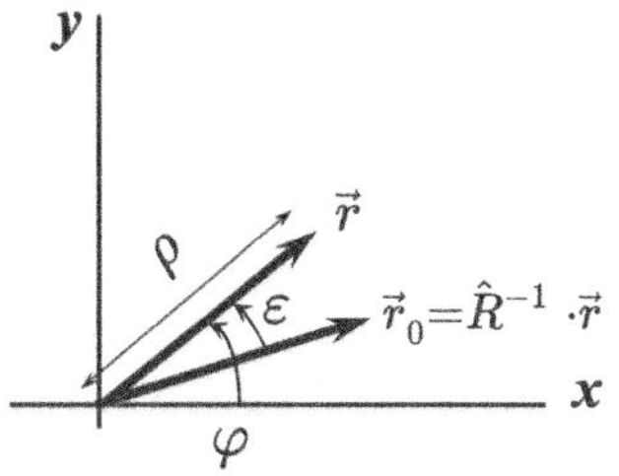

Fig. 7.1 Infinitesimal rotation around the z-axis

Using a Taylor expansion for the small angle $\varepsilon \ll 1$ as in Eq. (2.6), we find:

$$\begin{aligned} x_0 &= \rho \cdot \cos(\varphi - \varepsilon) \approx \rho \cdot [\cos\varphi - \sin\varphi \cdot (-\varepsilon)] \\ &= \rho \cdot \cos\varphi - \rho \cdot \sin\varphi \cdot (-\varepsilon) = x + \varepsilon \cdot y. \end{aligned}$$

Similarly, for y_0, we obtain:

$$x_0 = x + \varepsilon \cdot y, \quad y_0 = y - \varepsilon \cdot x, \quad z_0 = z.$$

Substituting these into Eq. (7.3), we have:

$$\left[\hat{R}_z(\varepsilon) \cdot \psi\right](\mathbf{r}) = \psi(x_0, y_0, z_0) = \psi(x + \varepsilon y, y - \varepsilon x, z).$$

Applying the Taylor expansion for the small terms εx and εy, we find:

$$\begin{aligned} \left[\hat{R}_z(\varepsilon) \cdot \psi\right](\mathbf{r}) &= \psi(x + \varepsilon y, y - \varepsilon x, z) \\ &= \psi(x, y, z) + \varepsilon y \cdot \frac{\partial}{\partial x}\psi - \varepsilon x \cdot \frac{\partial}{\partial y}\psi \\ &= \left[1 + \varepsilon\left(y\frac{\partial}{\partial x} - x\frac{\partial}{\partial y}\right)\right]\psi(x, y, z). \end{aligned} \tag{7.4}$$

At this point, using the angular momentum operator derived from the correspondence principle applied to the classical angular momentum $\mathbf{l} = \mathbf{r} \times \mathbf{p}$ (Sect. 4.1.2) as in Eq. (5.9), we have:

$$\hat{\mathbf{l}} = \mathbf{r} \times \hat{\mathbf{p}} = \mathbf{r} \times (-i\hbar\nabla) = -i\hbar(\mathbf{r} \times \nabla) \sim -i(\mathbf{r} \times \nabla), \tag{7.5}$$

where the differential operator appearing in Eq. (7.4) can be expressed as:

$$\hat{l}_z = -i\left(x\frac{\partial}{\partial y} - y\frac{\partial}{\partial x}\right).$$

Thus, the transformation $\hat{R}_z(\varepsilon)$ in Eq. (7.4) can be rewritten as:

$$\left[\hat{R}_z(\varepsilon)\cdot\psi\right](\mathbf{r}) = \left[1 - i\varepsilon\hat{l}_z\right]\psi(x, y, z).$$

The operator for an infinitesimal rotation around a general axis **n** can be derived using the same reasoning. For a direction cosine **n**, it is given by [1]:

$$\hat{R}_{\mathbf{n}}(\varepsilon) = \left[\hat{1} - i\varepsilon\cdot\mathbf{n}\cdot\hat{\mathbf{l}}\right]. \tag{7.6}$$

Furthermore, the angular momentum operator introduced in Eq. (7.5) satisfies the commutation relation:

$$\left[\hat{l}_\alpha, \hat{l}_\beta\right] = i\cdot\varepsilon_{\alpha\beta\gamma}\cdot\hat{l}_\gamma, \tag{7.7}$$

as demonstrated in Sect. 9.6.1.[3]

7.1.2 Spin of the Field as a Representation Matrix

Let us recall the discussion flow from the linear algebra section [Sect. 2.3.2] and proceed as follows.[4] In general, a field can be expanded as

$$\psi(\mathbf{x}) = \sum_{l=1}^{N} c_l\cdot\chi_l(\mathbf{x}), \tag{7.8}$$

where it is expressed in terms of an N-dimensional basis set.[5] The basis functions $\{\chi_j(\mathbf{x})\}_{j=1}^{N}$ transform among themselves under the coordinate transformation (6.7) as[6]

$$\hat{R}\cdot\chi_j(\mathbf{x}) = D_{\underline{k}j}\left(\hat{R}\right)\cdot\chi_{\underline{k}}(\mathbf{x}). \tag{7.9}$$

[3] Note that the equality on both sides of this equation refers to operators acting equivalently on $\psi(\mathbf{r})$. This algebraic structure is called the **Lie algebra** [2]. Here, $\varepsilon_{\alpha\beta\gamma}$ denotes the Eddington symbol introduced in Sect. 2.2.3.

[4] Refer to Fig. 1.2.

[5] The dimension N of the representation corresponds to $N = 1$ for scalar fields and $N = 3$ for vector fields.

[6] Pay attention to how the dummy index k (underlined) appears. This follows the standard basis transformation in vector spaces as in Eq. (6.7), where k is associated with the leading position in $D_{\underline{k}j}$. Representation theory of groups ultimately boils down to operations like block diagonalization of matrices. Experienced readers may wonder why, despite the distinct subject matter, the analysis ultimately reverts to linear algebra operations. This is because we assume that the transformed basis functions can still be expressed in terms of the original basis set. Hence, the ensuing mathematical treatment aligns with linear algebra.

The matrix $D_{ij}\left(\hat{R}\right)$ is referred to as the **representation matrix** for the transformation $\hat{R}$. Using the same logic as for Eq. (2.37) in Sect. 2.3.2, $a_{kj} = \left(\mathbf{e}_k, \hat{A}\cdot\mathbf{e}_j\right)$,[7] for orthonormal basis sets $\langle\chi_k|\chi_l\rangle = \delta_{kl}$, the representation matrix is expressed as

$$D_{ij}\left(\hat{R}\right) = \langle\chi_i|\,\hat{R}\,|\chi_j\rangle.$$

For the rotation operator acting on the field (7.6), let us write down its representation matrix:

$$\begin{aligned}\langle\chi_i(\mathbf{x})|\,\hat{R}_{\mathbf{n}}(\varepsilon)\,|\chi_j(\mathbf{x})\rangle &= \langle\chi_i(\mathbf{x})|\,\hat{1} - i\varepsilon\cdot\mathbf{n}\cdot\hat{\mathbf{l}}\,|\chi_j(\mathbf{x})\rangle \\ &= \hat{1}_{ij} - i\varepsilon\cdot\mathbf{n}\cdot\hat{\mathbf{L}}_{ij} \\ &=: R(\mathbf{n})_{ij} \end{aligned} \tag{7.10}$$

Thus, the angular momentum operator $\hat{\mathbf{l}}$ is represented as an $N\times N$ matrix,

$$\mathbf{L}_{ij} = \langle\chi_i|\hat{\mathbf{l}}|\chi_j\rangle, \quad (i,j) = 1,\cdots,N \tag{7.11}$$

This representation matrix of the angular momentum operator $\hat{\mathbf{L}}$ is referred to as the **spin of the field** [1].

When deriving the representation matrix for spatial rotations, it is customary to adopt the eigenfunctions of the angular momentum operator $\hat{l}_z$ as the expansion basis functions $|\chi_j\rangle$:

$$\hat{l}_z|m\rangle = m\cdot|m\rangle$$

If the maximum eigenvalue of $|m\rangle$ is M, the number of eigenstates is given by

$$|m\rangle = \{|-M\rangle, |-M+1\rangle, \cdots, |M-1\rangle, |M\rangle\}$$

resulting in $(2M+1)$ states, as derived in Sect. 9.6.2. This result is derived generally for algebras (Lie algebras) satisfying the commutation relations (7.7) by introducing the quantities

$$\hat{l}_\pm := \frac{1}{\sqrt{2}}\left(\hat{l}_x \pm i\cdot\hat{l}_y\right),$$

known as raising and lowering operators. Accordingly, the number of eigenstates $N = (2M+1)$ determines the **dimensionality of the field**, leading to classifications such as: "1-component fields (scalar fields) correspond to $M = 0$ (spin-0 fields)",

[7] Refer to the derivation flow of Eq. (2.37) in Sect. 2.3.2.

"3-component fields (ordinary vectors) correspond to $M = 1$ (spin-1 fields)", "5-component fields correspond to d-orbitals", "7-component fields correspond to f-orbitals", and so forth. Fields with dimension $N = (2M + 1)$ are referred to as **spin-M fields**.

7.1.3 Pauli Matrices as the Rotation Representation Matrix for Spinor Fields

In the relativistic electron theory discussed in the previous chapter, $|\psi_L^{(2)}\rangle$ appearing in (6.32) corresponds to a two-component field ($N = 2$), namely, a spin-1/2 field with $M = 1/2$. This field is referred to as a **spinor field**. In this case, the two eigenstates spanning the two-dimensional space of $\hat{l}_z|m\rangle = m \cdot |m\rangle$ are expressed as:

$$|m\rangle = \begin{pmatrix} |M\rangle \\ |M-1\rangle \end{pmatrix} = \begin{pmatrix} |+1/2\rangle \\ |-1/2\rangle \end{pmatrix} = \begin{pmatrix} |\uparrow\rangle \\ |\downarrow\rangle \end{pmatrix}.$$

By the way, when deriving the result $N = (2M + 1)$, the following relations are obtained [(9.44) 式]:

$$\hat{l}_-|M\rangle = \sqrt{M}|M-1\rangle, \quad \hat{l}_+|M-1\rangle = \sqrt{M}|M\rangle.$$

Using this rule, the representation matrix in (7.11) can be constructed as:

$$L_{ij}^{(\alpha)} = \begin{pmatrix} \langle\uparrow|\hat{l}_\alpha|\uparrow\rangle & \langle\uparrow|\hat{l}_\alpha|\downarrow\rangle \\ \langle\downarrow|\hat{l}_\alpha|\uparrow\rangle & \langle\downarrow|\hat{l}_\alpha|\downarrow\rangle \end{pmatrix}.$$

Explicitly, these matrices are given by:

$$L_{ij}^{(z)} = \frac{1}{2}\begin{pmatrix} 1 & 0 \\ 0 & -1 \end{pmatrix}, \quad L_{ij}^{(x)} = \frac{1}{2}\begin{pmatrix} 0 & 1 \\ 1 & 0 \end{pmatrix}, \quad L_{ij}^{(y)} = \frac{1}{2}\begin{pmatrix} 0 & -i \\ i & 0 \end{pmatrix}.$$

This can be demonstrated, as detailed in Sect. 9.6.3. These matrices are found to relate to the Pauli matrices σ_α introduced in the relativistic electron theory discussed in the previous chapter as:

$$L_{ij}^{(\alpha)} = \frac{1}{2}\sigma_\alpha. \tag{7.12}$$

In the previous chapter, specifically in (6.23), the Pauli matrices were introduced solely as a mathematical convenience to rewrite $E^2 = \cdots$ into the linear form $E = \cdots$. Here, it becomes evident that these Pauli matrices actually represent the transformation properties of the two-component field $|\psi_L^{(2)}\rangle$ under rotations, as defined by the representation matrix in (7.9).

7.2 Spinor Fields and Spin

The two-component quantity $|\psi_L^{(2)}\rangle$ appearing in the two-component Schrodinger equation (6.32) discussed in the previous chapter is a "two-component field (spinor field) defined on three-dimensional space". It can be expressed as:

$$|\psi_L^{(2)}\rangle = c_0 \cdot |\uparrow\rangle + c_1 \cdot |\downarrow\rangle.$$

Using the basis states $|\uparrow\rangle$ and $|\downarrow\rangle$, this can be rewritten as:

$$|\psi\rangle = c_0 \cdot |\uparrow\rangle + c_1 \cdot |\downarrow\rangle = c_0 \cdot |0\rangle + c_1 \cdot |1\rangle, \tag{7.13}$$

where $|\uparrow\rangle = |1\rangle$ and $|\downarrow\rangle = |0\rangle$. This representation allows it to be used as a quantum bit. As the standard textbooks say, this two-component state can be parameterized in the form of (1.2):

$$|\psi\rangle_{(\theta,\phi)} = \cos\frac{\theta}{2} \cdot |0\rangle + e^{i\phi} \sin\frac{\theta}{2} \cdot |1\rangle = \begin{pmatrix} \cos\frac{\theta}{2} \\ e^{i\phi} \sin\frac{\theta}{2} \end{pmatrix},$$

for example, by choosing $\theta = \pi/2$, the state becomes:

$$|\psi_{\pi/2,0}^{(2)}\rangle = \begin{pmatrix} \cos\frac{\pi}{4} \\ e^{i0} \sin\frac{\pi}{4} \end{pmatrix} = \frac{1}{\sqrt{2}}(|0\rangle + |1\rangle),$$

being described as "the two-component quantity is oriented in the x-direction in three-dimensional space". We shall now explore the meaning of this description.[8]

[8] Instead of merely stating, "This corresponds to the $SU(2)$ and $O(3)$ mapping", we aim to provide a more intuitive explanation to make the concept accessible even to beginners.

7.2.1 Direction of the Spinor Field

As described in (7.10) and based on (7.12), the transformation of a field under rotation $|\psi\rangle \to |\psi'\rangle$ for the two-component quantity in relativistic electron theory is expressed as

$$\begin{aligned}|\psi'\rangle_j &= R(\mathbf{n})_{jk} \cdot |\psi\rangle_k = \left(\hat{1}_{jk} - i\varepsilon \cdot \mathbf{n} \cdot \hat{\mathbf{L}}_{jk}\right) |\psi\rangle_k \\ &=: \hat{1}_{jk}|\psi\rangle_k - \frac{i\varepsilon}{2} \cdot \hat{U}(\mathbf{n})_{jk}\, |\psi\rangle_k. \end{aligned} \tag{7.14}$$

Incorporating the Pauli matrices into this, we find

$$\begin{aligned}\hat{U}(\mathbf{n})_{jk} &= \mathbf{n} \cdot \boldsymbol{\sigma} = \left[n_x \cdot \sigma_x + n_y \cdot \sigma_y + n_z \cdot \sigma_z\right] \\ &= \begin{pmatrix} n_z & n_x - in_y \\ n_x + in_y & -n_z \end{pmatrix}. \end{aligned} \tag{7.15}$$

The second term in (7.14), $\hat{U}(\mathbf{n}) \cdot |\psi\rangle$, generally produces a vector different from $|\psi\rangle$. As a result, $|\psi'\rangle$ represents a vector "oriented differently" compared to $|\psi\rangle$. Here, consider the familiar eigenvalue problem:

$$\hat{U}(\mathbf{n}) \cdot |\psi_{\mathbf{n}}\rangle = |\psi_{\mathbf{n}}\rangle. \tag{7.16}$$

The eigenstate satisfying this relationship remains proportional to itself under the action of $\hat{U}(\mathbf{n})$, meaning it does not "change direction". Restating this, the eigenstate $|\psi_{\mathbf{n}}\rangle$ is a state that does not change direction under rotations about the **n** axis. By the way, a vector "oriented along the **n** direction" does not change its direction under rotations about the **n** axis. Thus, $|\psi_{\mathbf{n}}\rangle$ can be interpreted as a "**state oriented in the n direction**".

When the spinor state in (1.2),

$$|\psi^{(2)}_{(\theta,\phi)}\rangle = \cos\frac{\theta}{2} \cdot |0\rangle + e^{i\phi}\sin\frac{\theta}{2} \cdot |1\rangle = \begin{pmatrix} \cos\frac{\theta}{2} \\ e^{i\phi}\sin\frac{\theta}{2} \end{pmatrix}$$

satisfies the condition for the eigenstate "aligned in the **n** direction" given in (7.16), the vector **n** and the parameters (θ, ϕ) are related as:

$$\mathbf{n} = \begin{pmatrix} n_x \\ n_y \\ n_z \end{pmatrix} = \begin{pmatrix} \sin\theta\cos\phi \\ \sin\theta\sin\phi \\ \cos\theta \end{pmatrix}, \tag{7.17}$$

as demonstrated in Sect. 9.6.4. Thus, $|\psi^{(2)}_{(\theta,\phi)}\rangle$ is interpreted as a state that does not change its orientation under rotation about the $\mathbf{n}$ axis, i.e., a state "oriented in the direction $\mathbf{n}$ with azimuthal angles (θ, ϕ) in three-dimensional space". The identification of $|\psi^{(2)}_{(\theta,\phi)}\rangle$ (two-component quantity) in (1.2) with the "spin variable $\boldsymbol{\sigma}$ (three-dimensional vector)" depicted on a Bloch sphere, as shown in Fig. 1.1a, is due to this reasoning.

7.2.2 *Controlling the Spinor Field in the Dirac Equation*

To summarize the content of Sect. 7.2.1, for the 2×2 Pauli matrices $\{\sigma_\alpha\}$, the eigenstate $|\psi_{\mathbf{n}}\rangle$ satisfying the relations

$$\hat{U}(\mathbf{n})_{jk} = \mathbf{n} \cdot \boldsymbol{\sigma}, \quad \hat{U}(\mathbf{n}) \cdot |\psi_{\mathbf{n}}\rangle = |\psi_{\mathbf{n}}\rangle$$

is interpreted as a quantity that represents a state that does not change its components under rotation about the $\mathbf{n}$ axis, i.e., a state "oriented in the $\mathbf{n}$ direction". This $\mathbf{n}$ is represented as the "spin state" on the Bloch sphere in three-dimensional space.

On the other hand, recall the discussion in Sect. 6.3.3: the "realized state" governed by the two-component Dirac equation in the non-relativistic limit is described as the "two-dimensional eigenvector of the matrix $-\mathbf{B} \cdot \boldsymbol{\sigma}$" [(6.33)]. Thus, $\mathbf{n}$ corresponds to $\mathbf{B}$. In other words, when a magnetic field $\mathbf{B}$ is applied to the electronic system, the spinor state of the system can be controlled to "point in the direction of $\mathbf{B}$ on the Bloch sphere".

When the direction of the applied magnetic field $\mathbf{B}$ is (θ, ϕ), the realized spinor state corresponds to the orientation specified by (7.17). If the applied magnetic field $\mathbf{B}$ is aligned in the x direction, then with $(\theta, \phi) = (\pi/2, 0)$, the resulting spinor state becomes:

$$|\psi^{(2)}_{\pi/2,0}\rangle = \begin{pmatrix} \cos\dfrac{\pi}{4} \\ e^{i0}\sin\dfrac{\pi}{4} \end{pmatrix} = \frac{1}{\sqrt{2}}(|0\rangle + |1\rangle) \tag{7.18}$$

This represents a "superposition state of $|0\rangle$ and $|1\rangle$".

7.3 Controlling Spin by Controlling Probability

Up to this point, we have deepened our understanding of spin states, particularly the "horizontal spin state". In the next chapter, this horizontal state plays a central role. Before moving forward, take a moment to review Sect. 5.2 ("Physical Quantities

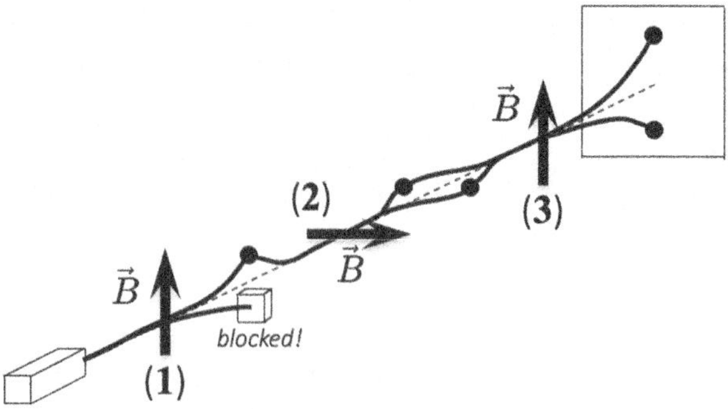

Fig. 7.2 Stern-Gerlach experiment. If we conceptualize spin like "a particle with an upward spin $|1\rangle$" and "a particle with a downward spin $|0\rangle$", the experimental facts depicted in the figure may seem perplexing. At the stage where particles enter magnet (2), downward spins are blocked. With the aforementioned conceptualization, one might wonder: "Why does the downward spin, which was once removed, reappear after passing through magnet (3)?!"

and Observation in Quantum Mechanics") by examining Fig. 7.2.[9] Whether one can view this experimental fact without any sense of discomfort serves as a test for properly understanding the "superposition state of $|0\rangle$ and $|1\rangle$".

When forming a picture of spin, if one adopts the perspective that there exist "particles with spin-up $|1\rangle$" and "particles with spin-down $|0\rangle$", then the experiment depicted in Fig. 7.2 may seem puzzlingdot At the stage where the spin-down particles are blocked by magnet (2), this perspective raises the question, "Why does the removed spin-down state reappear after passing through magnet (3)?!" Similarly, if one envisions that "spin particles oriented sideways exist" for a "spin state directed sideways", one might be confused as to "why the sideways-tilted spin in magnet (2) does not pass straight through the center in magnet (3) instead of being split into up and down directions".

Figure 7.2 illustrates the Stern-Gerlach experiment, a widely cited experimental fact when studying quantum mechanics. However, depending on the "teacher's narrative style", it might leave beginners with the impression of "What is so strange about this?" This can cast a shadow on their subsequent understanding. The lack of surprise often stems from the following reasoning: "If electrons have a magnetic moment called spin, it is only natural that applying a magnetic field would separate them up or down depending on the direction of the spin".

In the experimental setup depicted in Fig. 7.3,[10] the electron spins entering the magnet are expected to be oriented in random directions. From a classical perspective, one would anticipate a distribution like that shown in Fig. 7.3a. However, "remarkably", the actual experiment revealed that the electrons are deflected to only

9 Figures are created based on illustrations from Ref. [3].

10 Figures are created based on illustrations from Ref. [3].

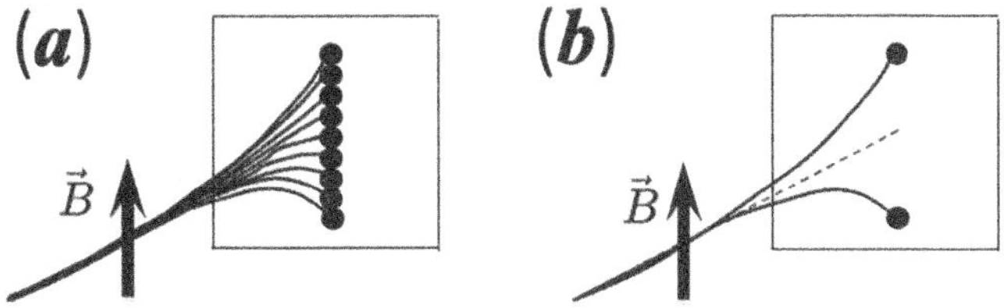

Fig. 7.3 The Stern-Gerlach experiment. The spins entering the magnet are expected to be oriented in random directions. From a classical perspective, one would anticipate a distribution like that shown in panel (**a**). However, "remarkably", the actual experiment revealed that the electrons are deflected to only two locations, as illustrated in panel (**b**)

two locations, as illustrated in Fig. 7.3b. This "finding" implies that a spin state in a general direction,

$$|\psi^{(2)}_{(\theta,\phi)}\rangle = \cos\frac{\theta}{2}\cdot|0\rangle + e^{i\phi}\sin\frac{\theta}{2}\cdot|1\rangle = \begin{pmatrix} \cos\frac{\theta}{2} \\ e^{i\phi}\sin\frac{\theta}{2} \end{pmatrix}$$

does not represent a "three-dimensional spin pointing in the direction (θ, ϕ)". Instead, it signifies a spin whose **internal degrees of freedom are probabilistically controlled**, such that $|0\rangle$ is observed with a probability $P_0 = |\cos(\theta/2)|^2$, and $|1\rangle$ is observed with a probability $P_1 = |\sin(\theta/2)|^2$.

To reiterate, the statement that "the applied magnetic field **B** polarizes the 'spin direction' depicted on the Bloch sphere" does not mean that the "three-dimensional spin direction" is being polarized. Instead, it signifies that the **probabilities of observing $|0\rangle$ and $|1\rangle$ are being controlled**. The representation on the Bloch sphere illustrates the "distribution of observation probabilities" in a convenient three-dimensional orientation. Please reaffirm this key interpretation.

With the above understanding, let us once again examine Fig. 7.2. After passing through Magnet (1) and "blocking the downward spin", the spin state of the electron is controlled to $|1\rangle$. Subsequently, upon passing through Magnet (2), the state is changed to

$$|\psi\rangle = \frac{1}{\sqrt{2}}[|0\rangle \pm |1\rangle] .$$

Therefore, at the location after passing through Magnet (3), $|0\rangle$ and $|1\rangle$ appear with equal probabilities.

References

1. Rose ME (2013) Elementary theory of angular momentum. Dover, New York. ISBN: 978-0486788791
2. Georgi H (1999) Lie algebras in particle physics: from isospin to unified theories. CRC Press, Boca Raton. ISBN: 978-0738202334
3. Coecke B, Kissinger A (2017) Picturing quantum processes. Cambridge University Press, Cambridge. ISBN: 978-1107104228

Chapter 8
Quantum Annealing

As the goal of this book, we will provide an overview of the principles of quantum annealing, building upon the concepts developed so far. We formulate combinatorial optimization problems as the search for the minimum of a potential landscape, and explain the idea of leveraging quantum tunneling to approach the solution (Sect. 8.1). This concept is concretely implemented in terms of finding the most stable configuration of a spin Hamiltonian (Sect. 8.2). To clarify the implementation, we present a detailed discussion of a specific example, formulating a path-finding problem (Sect. 8.3). Finally, as a supplement, we briefly explain how entanglement states emerge within quantum computation (Sect. 8.4).

8.1 Annealing in Optimization Problems

8.1.1 Optimization Problems and Local Minima

The **optimization problem** is traditionally understood by formulating it in terms of **minimum search**: treating various adjustable conditions as variables, the degree of conformity to the conditions is quantitatively expressed as a multivariable function. Choosing the input variables to minimize this evaluation function (taking its negative) corresponds to the act of "obtaining the optimal conditions". Plotting the evaluation function on the vertical axis over a plane where the horizontal axes represent the various adjustable conditions, a surface is drawn. Searching for the point that gives the minimum value on this functional surface corresponds to achieving the optimization problem (Fig. 8.1). Borrowing terminology familiar in physics, this can be understood as "the particle rolling down toward the state with the lowest energy, with the vertical axis representing the potential energy value".

R. Maezono, *Essentials for Deeper Understanding of Quantum Computing*,
https://doi.org/10.1007/978-981-96-5646-2_8

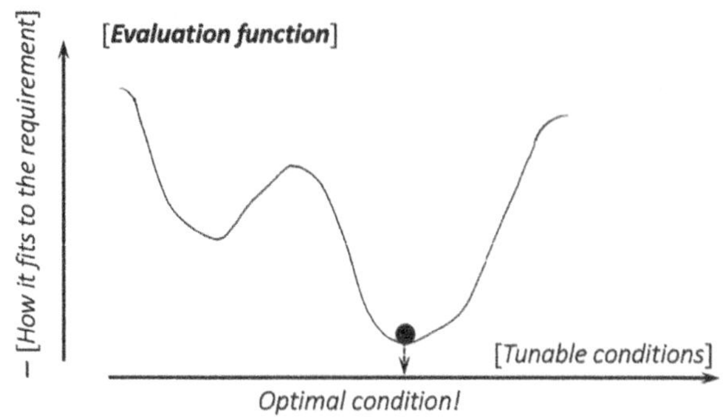

Fig. 8.1 Formulation of the optimization problem. The degree of conformity to the conditions is quantitatively expressed as a multivariable function with various adjustable conditions treated as variables. Searching for the point that gives the minimum value on the functional surface, such that the evaluation function (or its negative) is minimized, corresponds to achieving the optimization problem. Borrowing terminology familiar in physics, this can be understood as "the particle rolling down toward the state with the lowest energy, with the vertical axis representing the potential energy value"

The most elementary method for solving the problem of finding the minimum value of a function is to choose an initial position of the search point arbitrarily, evaluate the gradient of the function at that point, and descend along the steepest gradient direction (steepest descent method). However, if there are **local minima** scattered throughout and the minimum is not unique, such "methods based on gradients at local points" often result in getting trapped in a local minimum, preventing the discovery of the true optimal solution [global minimum]. The specific local minimum reached ultimately depends on the choice of the initial position.

In the physical analogy where the vertical axis represents potential energy, an effective approach is not to "update the state based on the gradient at the point of interest", but rather to "excite the search point to a higher-energy state (a point with a higher vertical axis value) temporarily and then gradually bring it down to a lower-energy state". This global minimum search strategy is metaphorically described as "heating the system to add energy, followed by **annealing** to bring it into the lowest energy state", and is referred to as 'simulated annealing'.

8.1.2 Inspiration for Quantum Annealing

In quantum phenomena, we mentioned that "state transitions that are not possible in classical theory" [virtual transition] can occur (Sect. 5.3.2). Utilizing such state

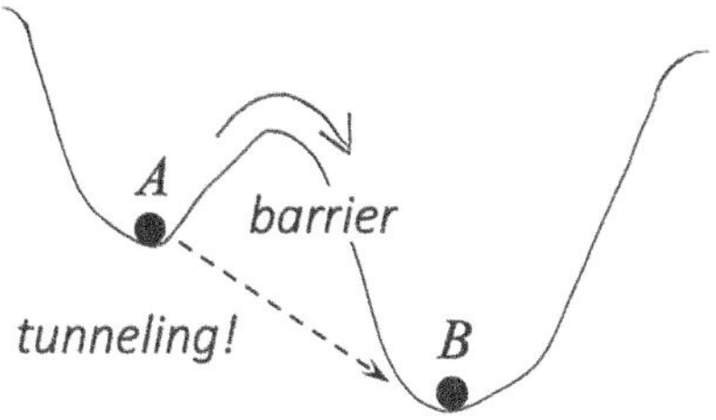

Fig. 8.2 In quantum phenomena, transitions that overcome "insurmountable energy barriers", which are not possible in classical theory, become feasible. This can be understood as tunneling through the barrier

transitions to overcome "insurmountable energy barriers" can be understood as **tunneling** through the barrier.

In the classical domain, a search point located in a valley of a multi-valley potential with local minima cannot escape from that valley unless sufficient energy is provided to overcome the valley. In contrast, when quantum mechanical virtual state transitions become possible, the search point can tunnel through and reach another valley (Fig. 8.2).

The reason the real world appears to be in a classical state is that it exists in an "extreme situation where quantum mechanical properties are not observable".[1] With advancements in nanotechnology, enabling conditions such as ultrafine scales or low temperatures to be achieved, the following can be realized:

- Ultrafine... Scales become comparable to the wavelength of the quantum mechanical wave ψ, and diffraction effects emerge.
- Low temperatures... The phase of the quantum mechanical wave ψ remains coherent without being disturbed by thermal effects.

Thus, a "stage where quantum virtual transitions are possible" can be realized. Quantum mechanical phenomena were described as "phenomena where the diffraction of the quantum mechanical wave ψ becomes prominent" (Sect. 5.1). For diffraction to occur, the phase of the wave must be clearly defined (Sect. 9.1.1). This clarity of phase is referred to as **coherence**.[2] When thermal effects disturb phase coherence to the point that it cannot be maintained, the transitions associated with quantum phenomena, such as virtual transitions, disappear. This can be understood as an image of "losing connectivity" due to the loss of phase correlation caused by thermal disturbances.

[1] As mentioned, the "tails" of quantum mechanical properties were discovered from the late nineteenth to the early twentieth century, leading to the revelation of the quantum mechanical governing laws previously unknown (Sect. 5.1).

[2] Laser light has high coherence, which is why it readily produces diffraction. On the other hand, natural light from sources like light bulbs has low coherence, making it difficult to observe diffraction without carefully narrowing the slits, as in Young's experiment.

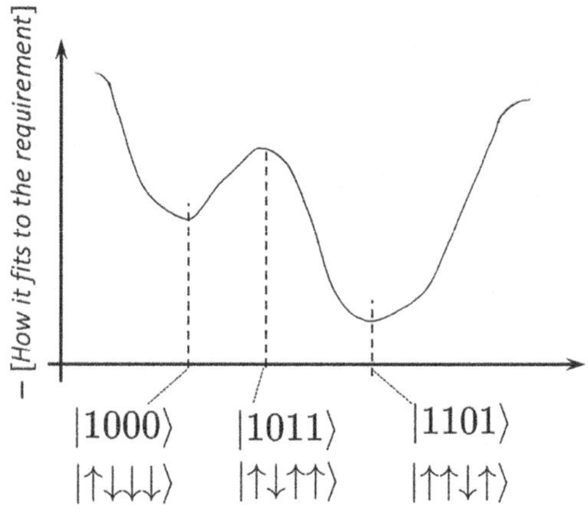

Fig. 8.3 Multi-valley potential and spin configuration. In quantum annealing, "states possible for the system" are represented by labeling them with binary numbers $|0000\rangle$, $|0001\rangle, \cdots$. Using $|1\rangle = |\uparrow\rangle$ and $|0\rangle = |\downarrow\rangle$, the horizontal axis values can be interpreted as corresponding to spin configurations

8.2 Overview of the Principles of Quantum Annealing

8.2.1 Strategy for Mapping Combinatorial Search Problems to Spin Models

In plots that represent "multi-valley structures in annealing", the horizontal axis corresponds to the "states possible for the system", which are the targets for optimal solution searches. In quantum annealing, these states are represented using **binary labeling**. This allows the horizontal axis values to be expressed as spin configurations such as $|1\rangle = |\uparrow\rangle$ and $|0\rangle = |\downarrow\rangle$, with states like $|0000\rangle$, $|0001\rangle, \cdots$ corresponding to spin arrangements (Fig. 8.3).

Now, the state of the electron spin given by Eq. (7.13)

$$|\psi_j\rangle = c_0 \cdot |\uparrow\rangle + c_1 \cdot |\downarrow\rangle$$

is a "quantum mechanical superposition state", where "upon observation, either $|\uparrow\rangle$ or $|\downarrow\rangle$ is observed exclusively".[3] Thus, the tensor product of these quantum bits $|\psi_j\rangle$,

$$\begin{aligned} |X\rangle &= |\psi_1\rangle \otimes |\psi_2\rangle \otimes \cdots \otimes |\psi_n\rangle \\ &= |\psi_1, \psi_2, \cdots, \psi_n\rangle \end{aligned} \tag{8.1}$$

represents a variable that takes on various horizontal-axis values in annealing, meaning "upon observation, it becomes $|0010\rangle$, or perhaps $|1010\rangle$". For example,

[3] The probability of observing either state is determined by the expansion coefficients c_0 and c_1.

considering the combination of three quantum bits,

$$\begin{aligned}
&|\psi_1\rangle \otimes |\psi_2\rangle \otimes |\psi_3\rangle \\
&= \left(c_0^{(1)} \cdot |0\rangle + c_1^{(1)} \cdot |1\rangle\right) \otimes \left(c_0^{(2)} \cdot |0\rangle + c_1^{(2)} \cdot |1\rangle\right) \otimes \left(c_0^{(3)} \cdot |0\rangle + c_1^{(3)} \cdot |1\rangle\right) \\
&= A_{000}|000\rangle + A_{001}|001\rangle + \cdots + A_{111}|111\rangle
\end{aligned} \tag{8.2}$$

is obtained,[4] and unless observed, the state $|X\rangle$ simultaneously realizes all 8 possible bit states from $|000\rangle$ to $|111\rangle$ in parallel.

A quantum bit $|\psi_j\rangle$ in a state oriented in the x-direction

$$|\psi_j\rangle = \begin{pmatrix} \cos\dfrac{\pi}{2} \\ e^{-i0}\sin\dfrac{\pi}{2} \end{pmatrix} \sim |0\rangle + |1\rangle \tag{8.3}$$

as described in Eq. (7.18), is a bit that "upon observation, is equally likely to be observed as $|0\rangle$ or $|1\rangle$ with a 50–50 probability". For the tensor product of these horizontally oriented spins $|\psi_j\rangle$, the coefficients $A_{001},\ A_{010}, \cdots$ in Eq. (8.2) are all equal, realizing a state where "all possible binary numbers on the horizontal axis are equally probable upon observation". This state is known as the **Hadamard state**. The Hadamard state is used as a fair initial estimate and serves as the starting point for solution searches.

The statement "some binary number $|X_0\rangle$ on the horizontal axis is the optimal solution" means that the vertical axis value is lowest at $|X_0\rangle$ (Fig. 8.3). How to prepare a function that provides such vertical axis values will be explained in the next section. By taking the Hadamard state as the initial estimate, a "superposition of all possibilities" is created. Using quantum mechanical tunneling, the state transitions to a lower energy state, ultimately reaching the optimal solution. This is the fundamental idea behind **quantum annealing**.

To solve combinatorial optimization problems using quantum annealing, it is necessary to reformulate the original problem into "all possibilities represented as spin configurations" (horizontal axis) and "energy gains or losses represented as spin configurations" (vertical axis). If the problem can be mapped onto such a **spin model**, it becomes possible to solve problems that are difficult for classical computing.

8.2.2 Spin Model

For the tensor product state $|X\rangle$ that represents a spin configuration,

$$|X\rangle = |\psi_1\rangle \cdots |\psi_i\rangle \cdots |\psi_j\rangle \cdots |\psi_n\rangle\,, \tag{8.4}$$

---

[4] Coefficients such as $A_{010} = c_0^{(1)} c_1^{(2)} c_0^{(3)}$, and states such as $|011\rangle = |0\rangle \otimes |1\rangle \otimes |1\rangle$, are abbreviated here.

we consider the Hamiltonian

$$\hat{H} = -J \cdot \boldsymbol{\sigma}_i \cdot \boldsymbol{\sigma}_j, \tag{8.5}$$

and examine the problem $\hat{H} \cdot |X\rangle = E|X\rangle$. The Pauli matrices $\boldsymbol{\sigma}_j$ act only on $|\psi\rangle_j$ in the tensor product state $|X\rangle$ given by Eq. (8.1). Therefore, the essential problem structure becomes

$$-J \cdot \boldsymbol{\sigma}_i \cdot \boldsymbol{\sigma}_j |\psi_i\rangle |\psi_j\rangle = E |\psi_i\rangle |\psi_j\rangle.$$

Thus, $|\psi_j\rangle$ is determined by

$$-J \cdot \boldsymbol{\sigma}_i \cdot \boldsymbol{\sigma}_j |\psi_j\rangle \propto E |\psi_j\rangle. \tag{8.6}$$

Recalling Sect. 7.2.2, the most stable state $|\psi_j\rangle$ satisfying the eigenvalue equation

$$\hat{H} \sim -\mathbf{B} \cdot \boldsymbol{\sigma}_j |\psi_j\rangle \propto E |\psi_j\rangle \tag{8.7}$$

corresponds to the orientation $\mathbf{B} = (\theta, \phi)$ on the Bloch sphere, expressed as

$$|\psi^{(2)}_{(\theta,\phi)}\rangle = \cos\frac{\theta}{2} \cdot |0\rangle + e^{i\phi} \sin\frac{\theta}{2} \cdot |1\rangle = \begin{pmatrix} \cos\frac{\theta}{2} \\ e^{i\phi} \sin\frac{\theta}{2} \end{pmatrix}.$$

Comparing Eq. (8.7) with Eq. (8.6), we see that $\mathbf{B} = -J \cdot \boldsymbol{\sigma}_i$. Thus, if $\boldsymbol{\sigma}_i$ is fixed in a given direction, the quantum bit $|\psi\rangle_j$ stabilizes in a state "oriented in the direction of $\boldsymbol{\sigma}_i$".

The above discussion holds equally well if the roles of $\boldsymbol{\sigma}_i$ and $\boldsymbol{\sigma}_j$ are interchanged. Thus, from $\hat{H} = -J \cdot \boldsymbol{\sigma}_i \cdot \boldsymbol{\sigma}_j$, we see that $|\psi\rangle_i$ and $|\psi\rangle_j$ tend to align in the same direction. In many standard texts on materials science or physics, this is explained as "$\hat{H} = -J \cdot \boldsymbol{\sigma}_i \cdot \boldsymbol{\sigma}_j$ maximizes the inner product when $\boldsymbol{\sigma}_i$ and $\boldsymbol{\sigma}_j$ are parallel, minimizing energy. Therefore, the electronic states at sites i and j align in the same direction". This explanation simplifies the description by **identifying** the state $|\psi_{i,j}\rangle$ with the orientation of $\boldsymbol{\sigma}_{i,j}$.[5] The parameter J corresponds to the energy required to flip a spin and is referred to as the **exchange interaction**.

[5] Many researchers in material-based science intuitively understand this, but when asked by students, "Is $\boldsymbol{\sigma}_i$ an electronic state? Wasn't it a Pauli matrix?", they often struggle to provide a clear explanation.

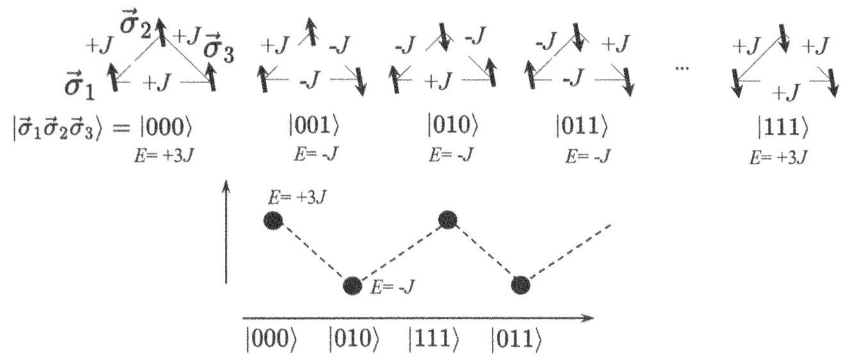

Fig. 8.4 The energy values given by spin configurations with frustration are expressed as a multi-valley state

8.2.3 Introduction of a Transverse Field

Next, we consider a Hamiltonian with site-dependence[6] for the exchange interaction ij, given by

$$\hat{H}_0 = -\sum_{<ij>} J_{ij} \cdot \boldsymbol{\sigma}_i \cdot \boldsymbol{\sigma}_j. \tag{8.8}$$

The summation symbol $< ij >$ indicates "the sum over ij pairs". From the discussion in the previous section, the sign of the exchange interaction can be understood as follows: if $J_{ij} > 0$, the ij pair tends to align parallel (ferromagnetic); if $J_{ij} < 0$, the ij pair tends to align anti-parallel (antiferromagnetic).

In this case, depending on the distribution of $\{J_{ij}\}$, it is possible to describe a multi-valley situation where "several locally stable spin configurations exist". To illustrate this schematically, let us consider a triangular lattice as shown in Fig. 8.4. Now, let us take the sign of all exchange interactions as $J_{ij} < 0$ and examine a spin model where "all adjacent spins want to align anti-parallel". In such a triangular lattice, it is impossible to align all pairs as anti-parallel,[7] resulting in several configuration patterns forming multiple local minima with equal energy (Fig. 8.4).[8] Here, we have explained the emergence of a multi-valley structure using a 3-bit configuration model, but it should be evident that using a configuration model with more sites (bits) allows for the representation of even more diverse multi-valley structures.

[6] This refers to the site where the quantum spins are located.

[7] This is referred to as a **frustration state**.

[8] These are referred to as multiple states with **degeneracy**.

Here, we consider adding a term describing a **transverse magnetic field** in the x direction

$$\hat{H}_B = -\mathbf{B} \cdot \sum_j \sigma_j, \quad \mathbf{B} = B \cdot \mathbf{e_x},$$

to Eq. (8.8), resulting in[9]

$$\begin{aligned} \hat{H} &= \hat{H}_0 + \hat{H}_B \\ &= -\sum_{<ij>} J_{ij} \cdot \sigma_i^z \cdot \sigma_j^z - B \sum_j \sigma_j^x. \end{aligned} \tag{8.9}$$

When the applied magnetic field B is strong, $\hat{H}_B$ dominates, and as described in Sect. 6.3.3, each bit stabilizes in a state "oriented in the x direction":

$$|\psi\rangle \sim |0\rangle + |1\rangle,$$

as given in Eq. (8.3). The tensor product constructed from this superposition state, $|X\rangle = |\psi_1\rangle|\psi_2\rangle \cdots |\psi_n\rangle$, forms the Hadamard state, where "all possibilities are equally realized", as described in Eq. (8.3). The situation where the applied magnetic field B dominates corresponds to a scenario like Fig. 8.5(a), where "the original multi-valley potential is lifted, realizing a uniform solution.[10]" This corresponds to the "initial state of annealing", where the system is "heated to add energy, allowing exploration into all valleys".

8.3 Applications of Quantum Annealing

8.3.1 Solution Search via Quantum Annealing

Although specific examples will be explained in the next section, let us assume that the multi-valley structure of a combinatorial search problem we wish to solve can be expressed as a spin model $\{J_{ij}\}$. In this case, finding the spin configuration $|X\rangle_0 = |\psi_1\rangle|\psi_2\rangle \cdots |\psi_n\rangle$ that gives the lowest energy for this spin model corresponds to solving the original combinatorial search problem (Sect. 8.2.1).

In this process, we first increase the transverse magnetic field B in the model Hamiltonian [apply a transverse field], realizing a superposition of all possible solutions as shown in Fig. 8.5a [Hadamard state]. From this state, by gradually

[9] The direction in which spins align under $\hat{H}_0$ is taken as the z-axis, while the transverse direction orthogonal to it is taken as the x-axis.

[10] This means all horizontal axis values are equally likely to be optimal.

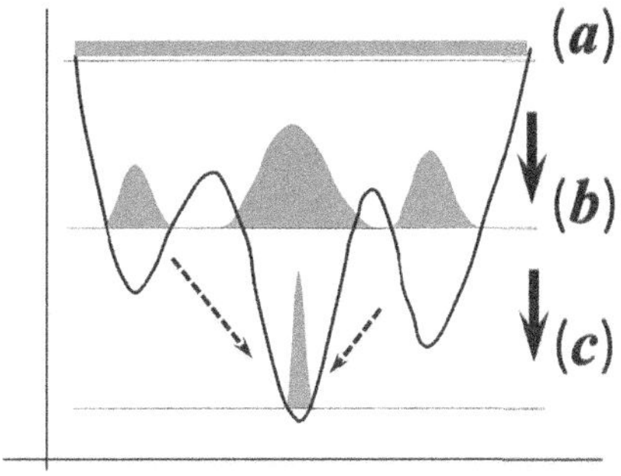

Fig. 8.5 In combinatorial optimization problems, we interpret the "set of possible solutions" as the horizontal axis and the evaluation function value as the vertical axis, drawing an analogy to energy values. A large transverse magnetic field B is initially applied, creating a superposition of all possible solutions as illustrated in (a) [Hadamard state]. Gradually weakening the transverse magnetic field [(a) $\rightarrow$ (b) $\rightarrow$ (c), as indicated by the arrows] allows tunneling effects to guide the solution space, progressively concentrating probability at the location corresponding to the minimum evaluation function value

weakening the transverse magnetic field [the arrow from Fig. 8.5(a) to (b)], the possibilities for the solution progressively concentrate around the location of the minimum value due to the tunneling effect. By doing so, quantum annealing overcomes the problem of "being trapped in a local minimum", which is challenging to avoid within the scope of classical computation.

The concept of quantum annealing was outlined using the expression "gradually reducing the transverse magnetic field". It is easy to imagine that the temporal efficiency in reaching the solution is influenced by the speed and pattern in which the transverse magnetic field is decreased. Fundamental theoretical research aimed at improving this efficiency has been actively conducted, and among these efforts, theoretical guarantees such as "always reaching the true ground state" have been established (the theory of quantum annealing) [1, 2].

8.3.2 Application to Combinatorial Optimization Problems

In quantum annealing, the framework becomes applicable if "practical combinatorial optimization problems can be mapped to a spin model...". To give a concrete sense of this concept, we will describe how a typical application example, the "route optimization problem", is handled.

As an example, let us consider a two-dimensional route optimization problem for a moving object on a 2D grid [3]. At time $t = t_k$, the position of the moving object is represented by a bright spot $|1\rangle$, while all other grid positions are represented by the value $|0\rangle$ [Fig. 1.1b]. The"0/1 pattern" sheet of the 2D grid, stacked incrementally over discrete time steps t_k, forms a "3D 0/1 array block", which represents a single route on the 2D grid (Fig. 8.6).

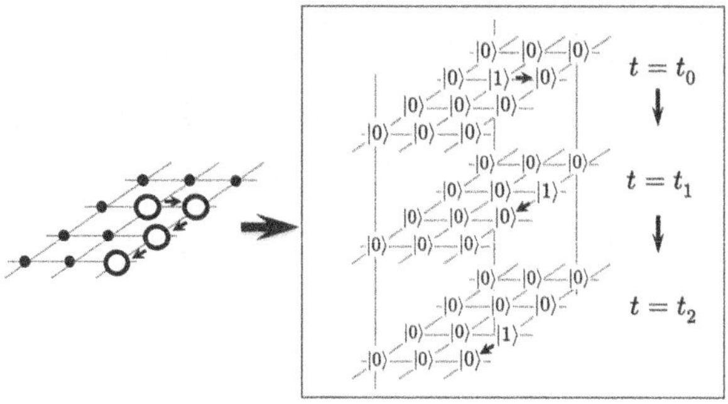

Fig. 8.6 Representation of routes in quantum annealing. A single route on a 2D grid (left) is represented as a 0/1 array block on a 3D grid

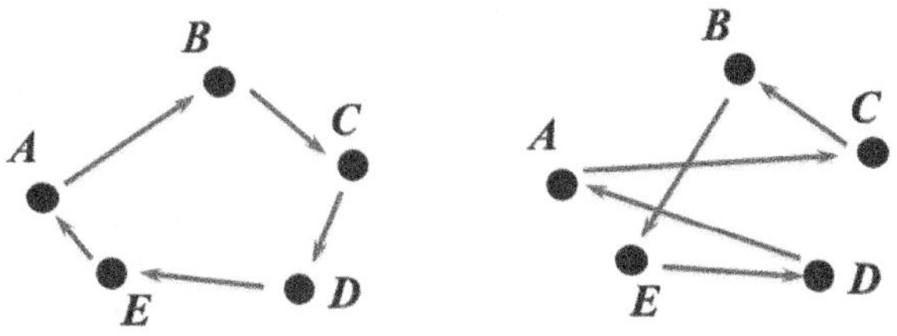

Fig. 8.7 Traveling salesman problem. Consider a route that visits all five locations, labeled $\alpha = A, B, \cdots, E$, exactly once and returns to the starting point. As illustrated, there are variations in the order in which the five locations can be visited. The problem is to find a route that minimizes the total travel distance

There are numerous possible routes on the 2D grid, each represented by a "3D block of 0/1 arrays". Finding a single optimal route, therefore, maps to the problem of identifying the optimal "3D block of 0/1 array patterns".

Having confirmed the approach of representing the search route as "bright spots $|0\rangle, |1\rangle$ on a 3D block", let us consider the following route optimization problem. Suppose there are five locations labeled $\alpha = A, B, \cdots, E$, and we aim to visit all of them exactly once and return to the starting point (the traveling salesman problem). As illustrated in Fig. 8.7, the order in which the five locations are visited can vary. The problem is then defined as finding a route that "minimizes the total travel distance". Let D_{AB} represent the travel distance (or, more generally, the cost) between points A and B. In this case, the total distance L depends on the chosen route, for example:

$$L_1 = D_{AB} + D_{BC} + D_{CD} + D_{DE} + D_{EA},$$

$$L_2 = D_{AC} + D_{CB} + D_{BE} + D_{ED} + D_{DA},$$

$$\ldots$$

Now, introducing a binary variable $s(\mathbf{r}, t)$, which corresponds to the bright spot representation:

$$s(\mathbf{r}, t) = \begin{cases} 1 \cdots |1\rangle \\ 0 \cdots |0\rangle \end{cases},$$

this variable represents whether the salesman is at a given location $\mathbf{r}$ at time t ($|1\rangle$) or not ($|0\rangle$). For this,

$$s(\mathbf{r}_\alpha, t)\, s\left(\mathbf{r}_\beta, t + \Delta t\right) = \begin{cases} 1 \cdots ([\alpha \to \beta] \text{ is a path for } [t, t + \Delta t]) \\ \qquad\qquad 0 \cdots \text{otherwise} \end{cases}.$$

The total distance L can then be written as:

$$L = \sum_t \sum_{\langle\alpha\beta\rangle} D_{\alpha\beta} \cdot s(\mathbf{r}_\alpha, t)\, s\left(\mathbf{r}_\beta, t + \Delta t\right).$$

For the summation over pairs $\langle\alpha\beta\rangle$, it is equivalent to include non-pair terms $(\alpha\beta)$ since $s(\mathbf{r}_\alpha, t)\, s\left(\mathbf{r}_\beta, t + \Delta t\right)$ will be zero in those cases and contribute nothing. Thus, the pair condition can be relaxed, and the expression becomes:

$$L = \sum_t \sum_{\alpha,\beta} D_{\alpha\beta} \cdot s(\mathbf{r}_\alpha, t)\, s\left(\mathbf{r}_\beta, t + \Delta t\right).$$

Here, to bring the notation closer to the appearance of spin model equations, we simplify the expressions. If the i-th time step is represented as "$t \to i$", then $(t + \Delta t) \to (i + 1)$. For positions, replacing $\mathbf{r}_\alpha = \alpha$ with site indices, we can write $s(\mathbf{r}_\alpha, t) \to s_{\alpha,i}$. Thus, the total distance L is rewritten as:

$$L = \sum_i \sum_{\alpha,\beta} D_{\alpha\beta} \cdot s_{\alpha,i} s_{\beta,i+1}. \tag{8.10}$$

For the bright spot states on a site,

$$s_{\alpha,i} = \begin{cases} 1 \cdots |1\rangle \\ 0 \cdots |0\rangle \end{cases},$$

the Pauli matrices act as:

$$\hat{\sigma}^{(z)}_{\alpha,i} |1\rangle = 1 \cdot |1\rangle, \quad \hat{\sigma}^{(z)}_{\alpha,i} |0\rangle = 0 \cdot |0\rangle,$$

essentially "picking up 0 or 1". For the tensor product of spin variables $|X\rangle$, as in Eq. (8.4), corresponding to whether the salesman is present ($|1\rangle$) or absent ($|0\rangle$) at each site, the Hamiltonian operator:

$$\hat{H}_0 = \sum_i \sum_{\alpha,\beta} D_{\alpha\beta} \cdot \hat{\sigma}^{(z)}_{\alpha,i} \hat{\sigma}^{(z)}_{\beta,i+1}$$

acts as:

$$\hat{H}_0 \cdot |X\rangle = L \cdot |X\rangle, \tag{8.11}$$

returning the total distance L as its eigenvalue. Indeed, $\hat{H}_0$ takes the form of a **spin model**, as seen in Eq. (8.8).

In the case of the traveling salesman problem, optimization must also consider the **constraints** that "the traveler is at only one location at any given time (single traveler)" and "each location is visited only once (single visit)". First, let us consider the condition that "at a fixed time i, the traveler is at only one location". This corresponds to the requirement that the sum of the bright spot variables $s_{\alpha,i}$ over all sites α for a fixed i equals 1. That is:

$$\sum_\alpha s_{\alpha,i} = 1. \tag{8.12}$$

Combining this with Eq. (8.10) for L, we define:

$$\tilde{L} = L + \lambda \cdot \left(\sum_\alpha s_{\alpha,i} - 1 \right)^2.$$

If the constraint in Eq. (8.12) is not satisfied, the λ term increases $\tilde{L}$ by a non-zero squared value. By choosing a large value for λ, this penalty term has a significant impact, ensuring that "violating the constraint leads to a higher evaluation function, resulting in a worse score". As a result, the optimization of $\tilde{L}$ proceeds in a way that satisfies the constraint. This method of incorporating constraints is referred to as the **penalty term** method.

The penalty term in the above $\tilde{L}$ represents the "single traveler condition at some specific time i". This condition must be satisfied for all times $i = 1, \cdots, N$. To reflect this, the penalty term can be incorporated as:

$$\begin{aligned} \tilde{L} &= L + \lambda \left[\left(\sum_\alpha s_{\alpha,1} - 1 \right)^2 + \left(\sum_\alpha s_{\alpha,2} - 1 \right)^2 + \cdots \right] \\ &= L + \lambda \sum_i \left(\sum_\alpha s_{\alpha,i} - 1 \right)^2. \end{aligned}$$

Next, let us consider the constraint that "the traveler visits each location α only once (single visit)". This means that for a fixed site α, the sum of the bright spot variables $s_{\alpha,i}$ over all times i equals 1. That is:

$$\sum_{i=1}^{N} s_{\alpha,i} = 1.$$

This can be expressed as a constraint that must hold for all sites. Using the same approach as for the single traveler condition, the constraint can be reflected with a penalty term as:

$$\tilde{\tilde{L}} = L + \lambda \sum_{\alpha} \left(\sum_{i=1}^{N} s_{\alpha,i} - 1 \right)^2.$$

From the above, the route search problem can be formulated as minimizing the evaluation function L_C:[11]

$$L_C = \sum_{\alpha,\beta} \sum_{i=1}^{N} D_{\alpha,\beta} \cdot s_{\alpha,i} \cdot s_{\beta,i+1} + \lambda_1 \sum_{\alpha} \left(\sum_{i=1}^{N} s_{\alpha,i} - 1 \right)^2 + \lambda_2 \sum_{i=1}^{N} \left(\sum_{\alpha} s_{\alpha,i} - 1 \right)^2.$$

Since this expression is quadratic in the bright spot variables $s_{\alpha,i}$, it can be compactly written as:

$$L_C = \sum_{\alpha,\beta} \sum_{i,j} J_{\alpha,\beta,i,j} \cdot s_{\alpha,i} \cdot s_{\beta,j}.$$

When given the parameter set $\{D_{\alpha,\beta}, \lambda_1, \lambda_2\}$, the corresponding $\{J_{\alpha,\beta,i,j}\}$ can be determined automatically using modern tools, making the process convenient and efficient.

Thus, as in Eq. (8.11), for this $\{J_{\alpha,\beta,i,j}\}$, the Hamiltonian operator:

$$\hat{H} = \sum_{\alpha,\beta} \sum_{i,j} J_{\alpha,\beta,i,j} \cdot \hat{\sigma}^z_{\alpha,i} \cdot \hat{\sigma}^z_{\beta,j}$$

acts as:

$$\hat{H} \cdot |X\rangle = L_C \cdot |X\rangle, \tag{8.13}$$

[11] The subscript 'C' in L_C represents the 'constraints'.

returning the optimization evaluation function L_C, which reflects the constraints, as its eigenvalue. With this, the route search problem is formulated as a problem of "finding the lowest eigenvalue of the spin Hamiltonian". From here, as described in Sects. 8.2.2 and 8.3.1, the optimal solution search can be realized using the operations of "generating a superposition state of all possibilities by applying a transverse magnetic field" and "relaxing the system by removing the transverse field, allowing tunneling".

8.4 Additional Remarks

Combinatorial optimization problems are "head-scratching issues" commonly encountered in daily life, such as in business optimization. Due to the high demand for solutions in the industrial sector, there was a period of significant enthusiasm when "quantum annealing, capable of solving such problems, became practical". However, as described above, the fundamental limitation is that "the problem must be expressible as a spin model". The most challenging aspect is developing a methodology to map real-world problems to a spin model $\{J_{ij}\}$. At the time of writing, this methodology is considered established only for a limited set of problems, such as the traveling salesman problem (TSP), partitioning problem, satisfiability problem (SAT), and clustering [2]. Nonetheless, even at the time of writing, there appear to be vigorous efforts to extend this horizon [4].

From the examples discussed in the previous section, it becomes evident that the following steps must be satisfied when performing "mapping to a spin model for applying quantum annealing":

- A spin Hamiltonian must be constructed such that the **relative order of magnitudes of the evaluation function** for the original problem are preserved.
- The spin Hamiltonian must be expressible within the quadratic range of spin variables.

A problem formulated within this "quadratic range" is referred to as QUBO (Quadratic Unconstrained Binary Optimization). The process of translating a problem into the spin model, as mentioned above, is often described as "whether the problem can be reduced to QUBO".

In the field of quantum information and communication technologies, including quantum computing, terms like **quantum entanglement** and **quantum entangled states** frequently appear. These are unique quantum concepts primarily utilized in applications such as quantum teleportation, long-distance communication, and quantum cryptography, and many readers might be curious about them. Although this book does not directly address this topic, we will briefly explain "what entanglement is" in the context of this book.

In quantum annealing and gate-based quantum computing,[12] the direct product $|X\rangle$ of multiple spin variables is utilized. For example, for a combination of three quantum bits,

$$\begin{aligned} &|\psi_1\rangle \otimes |\psi_2\rangle \otimes |\psi_3\rangle \\ &= \left(c_0^{(1)} \cdot |0\rangle + c_1^{(1)} \cdot |1\rangle\right) \otimes \left(c_0^{(2)} \cdot |0\rangle + c_1^{(2)} \cdot |1\rangle\right) \otimes \left(c_0^{(3)} \cdot |0\rangle + c_1^{(3)} \cdot |1\rangle\right) \\ &= A_{000}|000\rangle + A_{001}|001\rangle + \cdots + A_{111}|111\rangle. \end{aligned}$$

The initial value for computation, the Hadamard state, is a direct product of "superposed states for each individual bit". That is, $|X\rangle$ is in a state that "can be decomposed into a product of individual bits". However, during computation, each amplitude A_p changes from its initial state.[13] For instance, in a typical two-qubit example, the state might evolve into:

$$\frac{1}{\sqrt{2}}|0\rangle \otimes |0\rangle + \frac{1}{\sqrt{2}}|1\rangle \otimes |1\rangle = \frac{|00\rangle + |11\rangle}{\sqrt{2}}.$$

This state **cannot be decomposed** into a product of individual qubits, such as:

$$[\alpha_1|0\rangle + \beta_1|1\rangle] \otimes [\alpha_2|0\rangle + \beta_2|1\rangle].$$

No matter which α_j or β_j are used, this state cannot be expressed in this form. Such states, which cannot be "decoupled into a product of individual bits", are described as"**entangled states**". The phrase "quantum computation progresses by entangling states" is often used. This can be understood to mean that, depending on how amplitudes evolve during computation, such entangled states are traversed in the course of the calculation.

The fact that an entangled state cannot be "decoupled into a product of individual bits" implies that the quantum state of one bit is correlated with the quantum state of the other bit. If we imagine these two quantum bits being separated by a great distance, it leads to a conclusion seemingly incompatible with the classical local view, which claims that "events occurring far apart, with no means of influencing each other, should arise independently and locally". This has sparked significant controversy in the context of the interpretation of quantum mechanics [5]. The quantum computational techniques described in this book can be seen as actively leveraging the "strangeness of the interpretation problem". Similarly, quantum communication technologies, such as quantum teleportation, exploit the "peculiar non-locality of entangled states" to their advantage [6, 7].

[12] An overview is provided in Sect. 1.2.1. Readers who have reached this point may find it helpful to revisit this section for better understanding.

[13] In gate-based computing, this occurs artificially via gate operations, while in quantum annealing, it happens spontaneously due to tunneling as the transverse field is removed.

References

1. Das A, Chakrabarti BK (2005) Quantum annealing and related optimization methods. Springer, Heidelberg. ISBN: 978-3540279877
2. Tanaka S, Tamura R, Chakrabarti BK (2017) Quantum spin glasses, annealing and computation. Cambridge University Press, Cambridge. ISBN: 978-1107113190
3. Utimula K, Ichibha T, Prayogo G, Hongo K, Nakano K, Maezono R (2021) A Quantum annealing approach to ionic diffusion in solids. Sci Rep 11:7261. https://doi.org/10.1038/s41598-021-86274-3
4. Kitai K et al (2020) Designing metamaterials with quantum annealing and factorization machines. Phys Rev Res 2:013319, 2020. https://doi.org/10.1103/PhysRevResearch.2.013319
5. Weinberg S (2015) Lectures on quantum mechanics, 2nd edn. Cambridge University Press, Cambridge. ISBN: 978-1107111660
6. Nielsen MA, Chuang IL (2013) Quantum computation and quantum information. Cambridge University Press, Cambridge. ISBN: 978-1107619197
7. Mermin ND (2007) Quantum computer science: an introduction. Cambridge University Press, Cambridge. ISBN: 978-0521876582

Chapter 9
Appendix

This chapter contains appendices that include detailed derivations of mathematical expressions and more in-depth explanations of various concepts, which were set aside from the main text for later reference.

9.1 Supplemental for Chap. 1

9.1.1 Complex Amplitudes and Interference

Consider the sum of two complex numbers A_1 and A_2:

$$B = A_1 + A_2 \quad , \quad A_j \in \mathbb{C}. \tag{9.1}$$

Since complex numbers can be expressed in terms of amplitude and phase angle as $A_1 = |A_1| \, e^{i\phi_1}$, the sum B can be written as:

$$B = |A_1| \, e^{i\phi_1} + |A_2| \, e^{i\phi_2} = |B| \, e^{i\phi_B} .$$

The amplitude of the sum B is then evaluated as:

$$\begin{aligned} |B|^2 &= \left(|A_1| \, e^{i\phi_1} + |A_2| \, e^{i\phi_2}\right)^* \left(|A_1| \, e^{i\phi_1} + |A_2| \, e^{i\phi_2}\right) \\ &= |A_1|^2 + |A_2|^2 + |A_1| \, |A_2| \left(e^{-i\phi_1} e^{+i\phi_2} + e^{-i\phi_2} e^{+i\phi_1}\right) \\ &= |A_1|^2 + |A_2|^2 + 2 \, |A_1| \, |A_2| \cos\left(\phi_1 - \phi_2\right) . \end{aligned} \tag{9.2}$$

The important point here is that, in addition to the term corresponding to the sum of the amplitudes of the two waves, $\left(|A_1|^2 + |A_2|^2\right)$, there is an additional

R. Maezono, *Essentials for Deeper Understanding of Quantum Computing*,
https://doi.org/10.1007/978-981-96-5646-2_9

term involving $\cos(\phi_1 - \phi_2)$. If the phase angle difference between A_1 and A_2 is $(\phi_1 - \phi_2) = \pi$,[1] then $\cos(\phi_1 - \phi_2) = -1$, resulting in:

$$B_{\mathrm{OP}}^2 = \left(A_1^2 + A_2^2\right) - 2A_1A_2 = (A_1 - A_2)^2, \tag{9.3}$$

indicating that the amplitudes "cancel each other out."[2] Specifically, if the amplitudes of the two waves are equal ($A_1 = A_2$), they completely cancel out, and the resultant amplitude disappears.

To summarize, in the superposition of complex amplitudes as described by Eq. (9.1), the phase angle relationship between the components of the superposition causes **interference**, leading to a reduction or even cancellation of the resultant amplitude.

9.2 Supplementary Notes on Basic Mathematical Tools

9.2.1 Introduction of Inner Product

One of the key themes throughout this book is the formalization of "how the component representation changes when the coordinate axes of description are replaced". To manage this effectively, it is necessary to identify something that remains invariant, like the North Star. The length of a vector is independent of the choice of coordinate axes, and this invariance can be used as a starting point to construct quantities that are invariant under changes in coordinate systems. This leads us to the concept of the inner product of vectors.

The length of a three-dimensional vector is given by

$$|\mathbf{a}|^2 = a_x^2 + a_y^2 + a_z^2 = \left[a_x \cdot a_x + a_y \cdot a_y + a_z \cdot a_z\right].$$

Extending this calculation, we define the operation

$$\mathbf{a} \cdot \mathbf{b} = a_x \cdot b_x + a_y \cdot b_y + a_z \cdot b_z,$$

and call it the inner product. Using the inner product, the length of a vector can be written as

$$|\mathbf{a}|^2 = \mathbf{a} \cdot \mathbf{a} \quad \therefore \quad |\mathbf{a}| = \sqrt{\mathbf{a} \cdot \mathbf{a}}. \tag{9.4}$$

---

[1] This is expressed as the phase angles being "out of phase" with each other.

[2] The subscript "OP" in B_{OP} stands for 'Out of Phase'.

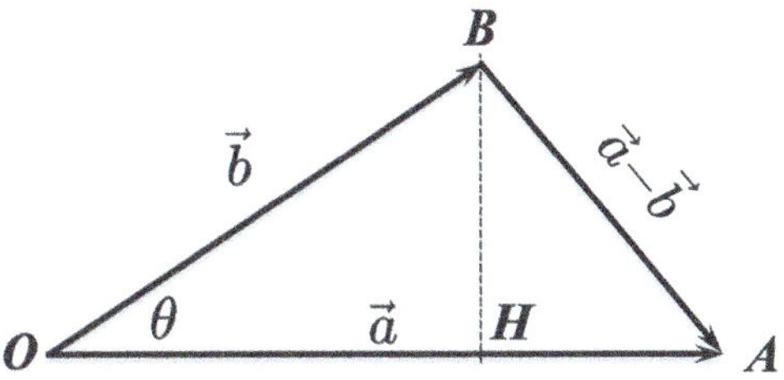

Fig. 9.1 Geometry for deriving the inner product formula. Let θ be the angle between **a** and **b**, and let H be the point where the perpendicular dropped from point B intersects the line OA

From this definition of the inner product, we can verify the distributive property:

$$\mathbf{a}\cdot(\mathbf{b}+\mathbf{c}) = a_x\,(b_x+c_x) + a_y\left(b_y+c_y\right) + a_z\,(b_z+c_z) = \mathbf{a}\cdot\mathbf{b}+\mathbf{a}\cdot\mathbf{c}. \quad (9.5)$$

Consider two vectors **a** and **b**, and let the angle between them be θ. Define the points as shown in Fig. 9.1. Using the Pythagorean theorem for the triangle ABH, we have:

$$AB^2 = HA^2 + HB^2.$$

The lengths of the non-hypotenuse sides can be expressed as:

$$HA = OA - OB\cdot\cos\theta, \quad , \quad HB = OB\cdot\sin\theta.$$

Substituting these into the Pythagorean theorem yields:[3]

$$\begin{aligned} AB^2 &= (OB\cdot\sin\theta)^2 + (OA - OB\cdot\cos\theta)^2 \\ &= (OB)^2 + (OA)^2 - 2OA\cdot OB\cdot\cos\theta. \end{aligned} \quad (9.6)$$

Rewriting this equation in terms of the lengths of vectors **a** and **b**, we get:

$$|\mathbf{a}-\mathbf{b}|^2 = |\mathbf{a}|^2 + |\mathbf{b}|^2 - 2\,|\mathbf{a}|\cdot|\mathbf{b}|\cdot\cos\theta.$$

On the other hand, using the relation between length and inner product given by Eq. (9.4) and the distributive property of the inner product (Eq. (9.5)), the left-hand side can also be expressed as:

$$(\mathbf{a}-\mathbf{b})\cdot(\mathbf{a}-\mathbf{b}) = |\mathbf{a}|^2 + |\mathbf{b}|^2 - 2\cdot(\mathbf{a}\cdot\mathbf{b})\,.$$

By comparing this with Eq. (9.7), we deduce:

$$\mathbf{a}\cdot\mathbf{b} = |\mathbf{a}|\cdot|\mathbf{b}|\cdot\cos\theta.$$

[3] It is important to ensure students clearly understand the Pythagorean theorem $\sin^2\theta + \cos^2\theta = 1$, possibly with a diagram of the unit circle to reinforce their understanding of the angle θ.

For N-dimensional vectors, we define the inner product as:

$$\mathbf{a} \cdot \mathbf{b} = a_1 \cdot b_1 + a_2 \cdot b_2 + \cdots + a_N \cdot b_N.$$

The value computed using this algorithm remains "invariant under changes of the descriptive coordinate axes". This property will be discussed later in Sect. 6.2.1.

9.2.2 Laplace Expansion of Determinants

First, let us confirm the equation:

$$A = \begin{vmatrix} 1 & 0 & \cdots & 0 \\ 0 & a_{22} & \cdots & a_{2N} \\ \vdots & \vdots & \ddots & \vdots \\ 0 & a_{N2} & \cdots & a_{NN} \end{vmatrix} = \begin{vmatrix} a_{22} & \cdots & a_{2N} \\ \vdots & \ddots & \vdots \\ a_{N2} & \cdots & a_{NN} \end{vmatrix}. \tag{9.7}$$

The determinant is defined as:

$$|A| = \sum_{P_1} \sum_{P_2 \cdots P_N} (-)^P \cdot a_{1P_1} a_{2P_2} \cdots a_{NP_N}.$$

In Eq. (9.7), the condition $a_{1P_1} = 1 \cdot \delta_{P_1 1}$ implies:

$$|A| = \sum_{P_2 \cdots P_N} (-)^P \cdot a_{2P_2} \cdots a_{NP_N}.$$

Since the summation over P_1 includes only terms where $P_1 = 1$, the set $(P_2, \cdots, P_N)$ corresponds to permutations of $2 \sim N$ that exclude 1. This summation becomes the determinant on the right-hand side of Eq. (9.7), confirming the equation.

Next, we derive Eq. (2.65) from the main text:

$$\begin{aligned} c_j^{(m)} &= |A|_{\{a_{P_m,m} = \delta_{[P_m = j]}\}} \\ &= (-)^{(m+j)} \sum_{P_2, P_3, \cdots} (-)^{P_2, P_3, \cdots} a_{P_2,1} \cdot a_{P_3,2} \cdots \cdots a_{N,P_N} \\ &= (-)^{(m+j)} |A_{jm}|. \end{aligned}$$

The left-hand side can be expressed as:

$$c_j^{(m)} = |A|_{\{a_{P_m,m}=\delta_{[P_m=j]}\}} = |A|_{\{a_{j,m}=1\}} = \begin{vmatrix} a_{11} & \cdots & a_{1j} & \cdots & a_{1N} \\ \vdots & \ddots & & & \vdots \\ 0 & \cdots & 1 & \cdots & 0 \\ \vdots & & & \ddots & \vdots \\ a_{N1} & \cdots & a_{Nj} & & a_{NN} \end{vmatrix},$$

(where the m-th row is $[0, \ldots, 1, \ldots, 0]$).

Using the basic transformations in Eq. (2.62), the m-th row can be repeatedly swapped upward to the first row [introducing a factor $(-)^{m+1}$]. Similarly, the j-th column is repeatedly swapped to the first column [introducing a factor $(-)^{j+1}$]:

$$c_j^{(m)} = (-)^{(j+1)+(m+1)} \begin{vmatrix} 1 & 0 & \cdots & \overset{(j)}{\cdots} & 0 \\ a_{2j} & a_{11} & a_{12} & \overset{(j)}{\cdots} & a_{1N} \\ \vdots & a_{21} & \ddots & & 0 \\ & \vdots & & \ddots & \vdots \\ a_{Nj} & a_{N1} & \cdots & \overset{(j)}{\cdots} & a_{NN} \end{vmatrix},$$

(where the m-th row is removed from rows 2 and onward).

Ultimately, using Eq. (9.7), we obtain:

$$c_j^{(m)} = (-)^{j+m} |A_{mj}|.$$

9.2.3 *Supplementary Calculations for Deriving the Inverse Matrix*

From Eq. (2.69), we write:

$$|A| = B_{ml} \cdot a_{lm} = a_{lm} \cdot B_l^{(m;A)}$$

where $B_l^{(m;A)}$ represents the "cofactor excluding the m-th column of matrix A", adopting a notation that recalls this concept.

Now, consider the off-diagonal term $B_{ml}a_{lm}$ evaluated as:

$$B_{jl} \cdot a_{lm}$$

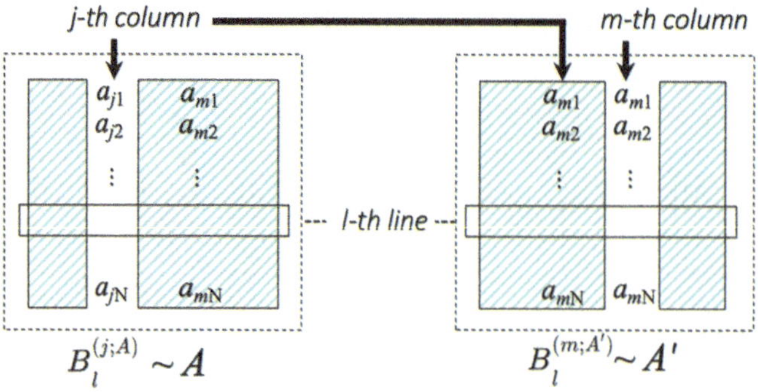

Fig. 9.2 $B_l^{(j;A)}$ is the "minor matrix determinant obtained by excluding the j-th column of A" (hatched region in the figure), with the l-th row further removed. When A' is formed by "overwriting the m-th column onto the j-th column" of A, the "minor matrix excluding the m-th column of A'" contains the same set of column elements as the minor obtained by excluding the j-th column of A, albeit in a different order. Therefore, $B_l^{(j;A)} = B_l^{(m;A')}$

Here, $B_l^{(j;A)}$ refers to the "minor matrix determinant obtained by excluding the j-th column of A" (represented by the hatched region in Fig. 9.2), with the l-th row further removed to form an $(N-1) \times (N-1)$ matrix. Now, consider a modified matrix A', formed by "overwriting the m-th column of A onto the j-th column". The "minor matrix obtained by excluding the m-th column of A'" (hatched region in the figure) contains the same set of column elements as the minor obtained by excluding the j-th column of A, albeit in a different order. Therefore, we have $B_l^{(j;A)} = B_l^{(m;A')}$, leading to:

$$B_{jl} \cdot a_{lm} = a_{lm} \cdot B_l^{(j;A)} = a_{ml} \cdot B_l^{(m;A')} = a'_{ml} \cdot B_l^{(m;A')}. \tag{9.8}$$

The reason for selecting the m-th column in constructing A' is that Eq. (9.8) aligns the indices (m, l), enabling the expression to match the form of a Laplace expansion for the determinant of A' along the m-th row and l-th column. As a result:[4]

$$B_{jl} \cdot a_{lm} = a_{lm} \cdot B_l^{(j;A)} = a'_{ml} \cdot B_l^{(m;A')} = |A'|.$$

Since A' is constructed so that its m-th and j-th rows are identical, its determinant satisfies $|A'| = 0$. Therefore:

$$B_{jl} \cdot a_{lm} = |A'| = 0.$$

Thus, the off-diagonal term $a_{ml} B_{lj}$ vanishes.

[4] The rightmost term holds because, by construction of A', $a_{ml} = a'_{ml}$.

9.2.4 Rotation Matrix

In high school, the rotation of a two-dimensional vector involves a matrix such as:

$$\hat{R}_\theta = \begin{pmatrix} \cos\theta & -\sin\theta \\ \sin\theta & \cos\theta \end{pmatrix}.$$

It is crucial to distinguish whether this represents: "(**1**) rotating a vector within the **same coordinate** system to obtain another vector", or "(**2**) rotating the coordinate system itself and observing how the components of the **same vector** change". The following discussion clarifies this distinction.

First, consider the case where"(**1**) a vector is rotated within the same coordinate system to yield another vector". A vector $\mathbf{v}$ represented as

$$\mathbf{v} = x \cdot \mathbf{e}_1 + y \cdot \mathbf{e}_2 = \begin{pmatrix} x \\ y \end{pmatrix},$$

undergoes a rotation by $\hat{R}_\theta$, resulting in $\mathbf{v} \to \hat{R}_\theta \cdot \mathbf{v}$. This transformation is expressed as:

$$\begin{pmatrix} x \\ y \end{pmatrix} \xrightarrow{\hat{R}_\theta} \begin{pmatrix} x' \\ y' \end{pmatrix} = \begin{pmatrix} \cos\theta & -\sin\theta \\ \sin\theta & \cos\theta \end{pmatrix} \begin{pmatrix} x \\ y \end{pmatrix},$$

as derived below. As shown in Fig. 9.3, the rotation operator $\hat{R}_\theta$ transforms:

$$\hat{R}_\theta \cdot \mathbf{e}_1 = \begin{pmatrix} \cos\theta \\ \sin\theta \end{pmatrix}, \quad \hat{R}_\theta \cdot \mathbf{e}_2 = \begin{pmatrix} -\sin\theta \\ \cos\theta \end{pmatrix}, \tag{9.9}$$

so that,

$$\mathbf{v} = x \cdot \mathbf{e}_1 + y \cdot \mathbf{e}_2 = \begin{pmatrix} x \\ y \end{pmatrix},$$

undergoes the following transformation:

$$\hat{R}_\theta \cdot \mathbf{v} = x \cdot \hat{R}_\theta \cdot \mathbf{e}_1 + y \cdot \hat{R}_\theta \cdot \mathbf{e}_2 = x \begin{pmatrix} \cos\theta \\ \sin\theta \end{pmatrix} + y \begin{pmatrix} -\sin\theta \\ \cos\theta \end{pmatrix}$$
$$= \begin{pmatrix} x \cdot \cos\theta - y \cdot \sin\theta \\ x \cdot \sin\theta + y \cdot \cos\theta \end{pmatrix} = \begin{pmatrix} \cos\theta & -\sin\theta \\ \sin\theta & \cos\theta \end{pmatrix} \begin{pmatrix} x \\ y \end{pmatrix}.$$

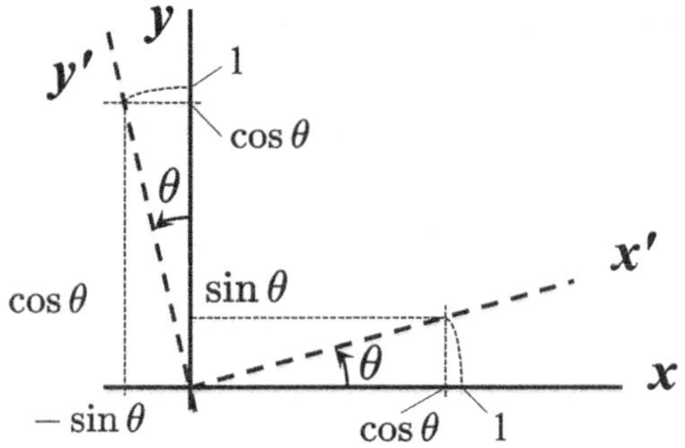

Fig. 9.3 Rotation in two-dimensional space

Thus,

$$\begin{pmatrix} x' \\ y' \end{pmatrix} = \hat{R}_\theta \begin{pmatrix} x \\ y \end{pmatrix} = \begin{pmatrix} \cos\theta & -\sin\theta \\ \sin\theta & \cos\theta \end{pmatrix} \begin{pmatrix} x \\ y \end{pmatrix},$$

and the rotation operator $\hat{R}_\theta$ for a rotation angle θ has the representation matrix:

$$\hat{R}_\theta = \begin{pmatrix} \cos\theta & -\sin\theta \\ \sin\theta & \cos\theta \end{pmatrix}.$$

Next, let us consider "**(2)** how the component representation of the same vector changes when the coordinate system is rotated". As shown in Fig. 9.3, when the coordinate system is rotated, the same vector is expressed differently in terms of the original coordinate system $\{\mathbf{e}_x, \mathbf{e}_y\}$ and the rotated coordinate system $\{\mathbf{e}'_x, \mathbf{e}'_y\}$ as:

$$\mathbf{v} = x \cdot \mathbf{e}_x + y \cdot \mathbf{e}_y = x' \cdot \mathbf{e}'_x + y' \cdot \mathbf{e}'_y.$$

These component representations are related by:

$$\begin{pmatrix} x \\ y \end{pmatrix} = \begin{pmatrix} \cos\theta & -\sin\theta \\ \sin\theta & \cos\theta \end{pmatrix} \begin{pmatrix} x' \\ y' \end{pmatrix}. \tag{9.10}$$

The derivation proceeds as follows: From Eq. (9.9),

$$\mathbf{e}'_x = \cos\theta \cdot \mathbf{e}_x + \sin\theta \cdot \mathbf{e}_y,$$
$$\mathbf{e}'_y = -\sin\theta \cdot \mathbf{e}_x + \cos\theta \cdot \mathbf{e}_y,$$

substitution into the expression for $\mathbf{v}$ yields:

$$\begin{aligned} \mathbf{v} &= x' \cdot [\cos\theta \cdot \mathbf{e}_x + \sin\theta \cdot \mathbf{e}_y] + y' \cdot [-\sin\theta \cdot \mathbf{e}_x + \cos\theta \cdot \mathbf{e}_y] \\ &= [x' \cdot \cos\theta - y' \cdot \sin\theta] \cdot \mathbf{e}_x + [x' \cdot \sin\theta + y' \cdot \cos\theta] \cdot \mathbf{e}_y. \end{aligned}$$

This yields Eq. (9.10), which serves as the **transformation rule between component representations** under a rotation of the coordinate system.

9.2.5 *Diagonalization of a Matrix*

We aim to find a basis system that satisfies the condition: "When the mapping $\hat{A}$ acts, there is no mixing between basis vectors", as described in Eq. (2.76):

$$\hat{A} \cdot \mathbf{e}'_j = \lambda^{(j)} \mathbf{e}'_j \tag{9.11}$$

Using this basis system, we consider expanding a general vector as follows:

$$\mathbf{x} = \sum_k x'_k \cdot \mathbf{e}'_k$$

For the same vector, whether expressed using the original orthogonal basis set $\{\mathbf{e}_j\}$ or the non-hybridizing basis set $\{\mathbf{e}'_j\}$, we write:

$$\mathbf{x} = \begin{pmatrix} x_1 \\ x_2 \\ \vdots \\ x_N \end{pmatrix}_{\{\mathbf{e}_j\}} = \begin{pmatrix} x'_1 \\ x'_2 \\ \vdots \\ x'_N \end{pmatrix}_{\{\mathbf{e}'_j\}}$$

$$\text{i.e.,} \quad \mathbf{x} = \sum_j x_j \cdot \mathbf{e}_j = \sum_j x'_j \cdot \mathbf{e}'_j \tag{9.12}$$

Assume the N solutions to the eigenvalue equation in Eq. (9.11) are:

$$\mathbf{e}'_1 = \begin{pmatrix} p_{11} \\ p_{21} \\ \vdots \\ p_{N1} \end{pmatrix}_{\{\mathbf{e}_j\}}, \quad \mathbf{e}'_2 = \begin{pmatrix} p_{12} \\ p_{22} \\ \vdots \\ p_{N2} \end{pmatrix}_{\{\mathbf{e}_j\}}, \quad \cdots, \quad \mathbf{e}'_N = \begin{pmatrix} p_{1N} \\ p_{2N} \\ \vdots \\ p_{NN} \end{pmatrix}_{\{\mathbf{e}_j\}} \tag{9.13}$$

This can be expressed as:[5]

$$\mathbf{e}'_j = \sum_l p_{lj} \cdot \mathbf{e}_l . \tag{9.14}$$

[5] Be cautious about the indices of p_{lj}. These indices are related to the basis vectors and must be assigned as shown in Eq. (2.48). Equation (9.14) needs to be established first. Upon rewriting, this corresponds to Eq. (9.13), which relates the components of $\mathbf{e}'_j$ to p_{ik} as in Eq. (9.13).

Substituting Eq. (9.14) into Eq. (9.12), we have:

$$\mathbf{x} = \sum_l x_l \cdot \mathbf{e}_l = \sum_l x'_l \cdot \mathbf{e}'_l = \sum_j x'_l \cdot \sum_m p_{ml} \cdot \mathbf{e}_m$$
$$= \sum_{l,m} p_{ml} \cdot x'_l \cdot \mathbf{e}_m$$

Renaming the dummy indices $m \to l$ and $l \to k$ for clarity, we get:

$$\mathbf{x} = \sum_l x_l \cdot \mathbf{e}_l = \sum_l \sum_k p_{lk} \cdot x'_k \cdot \mathbf{e}_l$$

By comparing coefficients of the basis $\mathbf{e}_l$, we obtain:

$$x_l = \sum_k p_{lk} \cdot x'_k \tag{9.15}$$

Explicitly written:[6]

$$\begin{pmatrix} x_1 \\ x_2 \\ \vdots \\ x_N \end{pmatrix} = \begin{pmatrix} p_{11} & p_{12} & \cdots & p_{1N} \\ p_{21} & \ddots & & \\ \vdots & & & \\ p_{N1} & \cdots & & p_{NN} \end{pmatrix} \begin{pmatrix} x'_1 \\ x'_2 \\ \vdots \\ x'_N \end{pmatrix} . \tag{9.16}$$

This establishes the relationship between component representations in Eq. (9.12).

Summarizing: when N eigenvectors are determined as in Eq. (9.13), they can be arranged as columns into a square matrix:

$$P = \left(\mathbf{e}'_1, \mathbf{e}'_2, \cdots, \mathbf{e}'_N\right) = \begin{pmatrix} p_{11} & p_{12} & \cdots & p_{1N} \\ p_{21} & \ddots & & \\ \vdots & & & \\ p_{N1} & \cdots & & p_{NN} \end{pmatrix} . \tag{9.17}$$

Using this matrix, the **transformation between representations in the new and old basis systems** can be written as:[7]

$$\mathbf{x} = P \cdot \mathbf{x}' \quad , \quad \therefore \quad \mathbf{x}' = P^{-1}\mathbf{x}. \tag{9.18}$$

[6] Although this resembles the transformation law for vectors in the same basis, in this context, it represents the transformation rules **between components** expressed in different basis systems. To avoid confusion, it is preferable to become familiar with component notation like Eq. (9.15) rather than vector notation as in Eq. (9.16).

[7] Note that this does not describe the transformation law for vectors within the same basis but instead represents the transformation rules between component representations in different basis systems.

Next, consider the matrix representation of a linear mapping $\hat{A}$ for vectors, expressed as:

$$\mathbf{y} = A \cdot \mathbf{x} \quad (\{\mathbf{e}_j\}\text{-frame}) \tag{9.19}$$

and in the $\{\mathbf{e}'_j\}$-frame as:[8]

$$\mathbf{y}' = A' \cdot \mathbf{x}' \quad \left(\{\mathbf{e}'_j\}\text{-frame}\right) \tag{9.20}$$

We derive the relationship between the representations A and A' using Eq. (9.18).

Substituting Eq. (9.18) into Eq. (9.20) yields:

$$\mathbf{y}' = A' \cdot P^{-1}\mathbf{x}.$$

Applying the rule in Eq. (9.18) to the left-hand side, we have:

$$P^{-1}\mathbf{y} = A' \cdot P^{-1}\mathbf{x}.$$

Multiplying both sides by P, we get:

$$P \cdot P^{-1}\mathbf{y} = P \cdot A'P^{-1}\mathbf{x} \quad , \quad \therefore \quad \mathbf{y} = \left(P \cdot A'P^{-1}\right)\mathbf{x}.$$

Comparing this with Eq. (9.19), we obtain:

$$A = P \cdot A'P^{-1} \quad , \quad \therefore \quad A' = P^{-1}AP. \tag{9.21}$$

This corresponds to the **transformation rule for matrix representations between new and old basis coordinates.**

In Sect. 2.3.9, we discussed that the matrix representation in the $\{\mathbf{e}'_j\}$-basis, where the eigenvectors are used for the expansion basis, is expected to be diagonal. This means A' should be diagonal. To verify this, substitute the expression of P from Eq. (9.17) into $A' = P^{-1}AP$:

$$\begin{aligned} A \cdot P &= A \cdot (\mathbf{e}'_1, \mathbf{e}'_2, \cdots, \mathbf{e}'_N) \\ &= (A\mathbf{e}'_1, A\mathbf{e}'_2, \cdots, A\mathbf{e}'_N) \\ &= \left(\lambda^{(1)}\mathbf{e}'_1, \lambda^{(2)}\mathbf{e}'_2, \cdots, \lambda^{(N)}\mathbf{e}'_N\right) \end{aligned}$$

[8] Here, $\hat{A}$ is an abstract operator. A and A' in Eqs. (9.19) and (9.20) are the matrix representations of the same operator $\hat{A}$ in different coordinate systems. They differ due to the change in basis and are denoted with distinct symbols.

$$= \left(\mathbf{e}'_1, \mathbf{e}'_2, \cdots, \mathbf{e}'_N\right) \begin{pmatrix} \lambda^{(1)} & 0 & \cdots & 0 \\ 0 & \lambda^{(2)} & & \\ \vdots & & \ddots & \\ 0 & & & \lambda^{(N)} \end{pmatrix} = PD,$$

where D is the diagonal matrix. Multiplying both sides by P^{-1}, we get:

$$P^{-1} \cdot AP = P^{-1} \cdot PD = D,$$
$$\therefore \quad A' = P^{-1} \cdot AP = D.$$

Thus, the matrix representation A' in the new coordinate system spanned by the eigenvectors is shown to be diagonal.

In this book, the determinant is defined and introduced as the "scaling factor" of a mapping (Sect. 2.3.5). Accordingly, it is expected, as stated in Sect. 2.3.9, that:

$$|A| = \prod_{j=1}^{N} \lambda^{(j)}. \tag{9.22}$$

This can be verified by applying the property:

$$\det[AB] = \det[A] \cdot \det[B]$$

to the relationship $A = P \cdot A' \cdot P^{-1}$:

$$|A| = \left|P \cdot A' \cdot P^{-1}\right| = |P| \cdot \left|A'\right| \cdot \left|P^{-1}\right| = \left|A'\right| = |D| = \prod_{j=1}^{N} \lambda^{(j)}.$$

Thus, this result can be directly confirmed.

9.2.6 Characteristic Equation of the Eigenvalue Problem

The eigenvalue equation introduced in Eq. (2.76) is expressed as:

$$A \cdot \mathbf{e}'_j = \lambda^{(j)} \mathbf{e}'_j \quad , \quad \therefore \quad (A - \lambda I) \cdot \mathbf{e}' = \mathbf{0},$$

where the problem is to solve this equation.[9] The above equation implies that the projection operator $(A - \lambda I)$ maps any vector to the zero vector. If $(A - \lambda I)$ has

[9] Here, $\lambda^{(j)} = \lambda$ and $\mathbf{e}'_j = \mathbf{e}'$ are used as representatives, dropping the index j.

an inverse matrix, multiplying both sides by the inverse yields:

$$\mathbf{e}' = (A - \lambda I)^{-1} \cdot \mathbf{0} = \mathbf{0},$$

which provides only the trivial solution $\mathbf{e}' = \mathbf{0}$. For a non-trivial solution, it is necessary that $(A - \lambda I)$ does not have an inverse matrix. According to the formula for the inverse matrix in Eq. (2.72):

$$A^{-1} = \frac{\mathrm{Cof}(A)}{|A|},$$

if the determinant of the matrix in the denominator is zero, the formula becomes undefined, and the inverse matrix does not exist. Thus, the requirement for $(A - \lambda I)$ is:

$$|A - \lambda I| = 0,$$

which is the condition to obtain a non-trivial $\mathbf{e}'$. This is referred to as the **characteristic equation**.

For an N-dimensional matrix A, the characteristic equation becomes an N-th degree polynomial equation in λ. According to the fundamental theorem of algebra, this equation generally has N solutions. In the context of the basis transformation from $\{\mathbf{e}_l\}$ to $\{\mathbf{e}'_l\}$, this corresponds to the statement that there are N basis vectors $\{\mathbf{e}'_l\}$ for the N basis vectors $\{\mathbf{e}_l\}$.

9.2.7 Introduction of Bra-ket Notation

Bra-ket notation is typically introduced in quantum mechanics courses, which leads many beginners to mistakenly believe it is specific to quantum systems. However, it is a convenient notation for handling function spaces and can be applied equally to classical and quantum systems when using basis set function expansions.[10]

In Sect. 2.4.4, we introduced bra-ket notation,[11] where a function $f(x)$ is expressed as:

$$f(x) = \langle x | f \rangle$$

[10] The emphasis on this point comes from the author's experience teaching in information science departments. Students not specializing in physics, despite having strong academic ability, often give up understanding bra-ket notation, assuming it to be an unfamiliar topic outside their expertise. What characterizes quantum systems is not this notation but the context where "physical states" are expressed as superpositions in the form of function expansions.

[11] In this book, bra-ket notation is introduced somewhat informally, but there is a more rigorous formalism [3]. For now, it is best to get used to the notation as presented here and later explore the "common pitfalls" pointed out in stricter formal treatments.

This notation can be interpreted as the "inner product of f and x", which **bewilders beginners** who wonder what "inner product with x" actually means. For now, simply accept this as a notation convention, since treating it like an inner product yields consistent results.[12] The above expression can be read as: "The projection of the abstract vector $|f\rangle$ onto the Euclidean vector space $|x\rangle$ is $\langle x|f\rangle$, which corresponds to the function $f(x)$."[13] Using this notation allows us to develop a framework analogous to the one constructed for Euclidean vector spaces, providing clearer insights [1].

Consider an expansion using an orthogonal basis set function[1, 2]:

$$f(x) = \sum_l f_l \cdot \chi_l(x) \quad , \quad \int dx \cdot \chi_j^*(x)\, \chi_l(x) = \delta_{kl}$$

The expansion coefficients $\{f_j\}$ are given by:

$$\int dx \cdot \chi_j^*(x) \cdot f(x) = \sum_l f_l \int dx \cdot \chi_j^*(x) \cdot \chi_l(x) = \sum_l f_l \cdot \delta_{jl} = f_j$$

$$\therefore \quad f_j = \int dx \cdot \chi_j^*(x) \cdot f(x) .$$

Using the above bra-ket notation, the basis set expansion can be expressed as:

$$\langle x|f\rangle = \sum_l f_l \cdot \langle x|\chi_l\rangle \quad , \quad f_j = \langle \chi_j|f\rangle . \tag{9.23}$$

If the basis set is complete, then by substituting the projection components f_j back into Eq. (9.23), the original $f(x) = \langle x|f\rangle$ can be reconstructed:

$$\langle x|f\rangle = \sum_l \langle \chi_l|f\rangle \cdot \langle x|\chi_l\rangle . \tag{9.24}$$

[12] The bra-ket $\langle f|g\rangle$ is initially explained as a way to represent an inner product. More precisely, it is introduced in the context of "linear functionals and dual spaces", where $\langle f|$ is a dual space element that maps a vector $|g\rangle$ to a scalar $I = \langle f|g\rangle$. $\psi(x)$ can be interpreted as a "linear functional that maps the element $\langle x|$ in a vector space to the complex number $\psi(x)$". This linear functional, combined with $\langle x|$, forms $\langle \psi|$, which is the meaning conveyed by the notation $\psi(x) = \langle x|\psi\rangle$.

[13] The idea of considering a function as an "abstract entity $|f\rangle$", with its method of representation chosen each time according to human convenience, is a more modern perspective. This way of thinking also underlies concepts such as analytic continuation in complex function theory[4].

Reordering the product on the right-hand side, we get:

$$\langle x|f\rangle = \sum_l \langle x|\chi_l\rangle \langle \chi_l|f\rangle = \langle x| \cdot \sum_l |\chi_l\rangle\langle \chi_l| \cdot |f\rangle.$$

This can be interpreted as inserting the identity operator $\langle x|f\rangle = \langle x| \cdot \hat{1} \cdot |f\rangle$, where the identity projection is given by:

$$\hat{1} = \sum_l |\chi_l\rangle\langle \chi_l|. \tag{9.25}$$

Since Eq. (9.25) is derived from the assumption of completeness, it is called the **completeness condition** of the basis set $\{|\chi_j\rangle\}$. By recalling this, inserting the completeness condition $\langle x|f\rangle = \langle x| \cdot \hat{1} \cdot |f\rangle$ enables you to formally derive the general basis function expansion in Eq. (9.24). This provides a convenient and systematic way to handle expansions in any complete basis set.

The completeness condition and identity projection in the case of a continuous system can be derived in a similar way:

$$f(x) = \int d\omega \cdot f(\omega) \cdot \chi_\omega(x).$$

This expression can be envisioned as a "Fourier spectrum $f(\omega)$" and "Fourier basis function $\chi_\omega(x)$". The spectral function $f(\omega)$ can be obtained by the projection $f(\omega) = \langle \chi_\omega|f\rangle$, resembling an inverse Fourier transform. Substituting this, we get:

$$\begin{aligned}\langle x|f\rangle &= \int d\omega \cdot \langle \chi_\omega|f\rangle \cdot \langle x|\chi_\omega\rangle \\ &= \int d\omega \cdot \langle x|\chi_\omega\rangle \langle \chi_\omega|f\rangle \\ &= \langle x| \cdot \int d\omega\, |\chi_\omega\rangle \langle \chi_\omega| \cdot |f\rangle.\end{aligned}$$

Comparing this with:

$$\langle x|f\rangle = \langle x| \cdot \hat{1} \cdot |f\rangle,$$

we obtain the identity projection expression:

$$\hat{1} = \int d\omega\, |\chi_\omega\rangle \langle \chi_\omega|$$

9.2.8 Some Expressions Using Bra-Ket Notation

The expression for the Fourier transform is:

$$f(x) = \frac{1}{\sqrt{2\pi}} \int dp \cdot e^{ipx} \cdot f(p).$$

On the other hand, using bra-ket notation, $f(x)$ can be expressed as:

$$f(x) = \langle x | f \rangle = \int dp \cdot \langle x | p \rangle \langle p | f \rangle = \int dp \cdot \langle x | p \rangle f(p).$$

By inserting the identity relation for $|p\rangle$:

$$\hat{1} = \int dp \, |p\rangle \langle p| \tag{9.26}$$

Comparing the two expressions, we get:

$$\langle x | p \rangle = \frac{1}{\sqrt{2\pi}} \cdot e^{ipx}. \tag{9.27}$$

Similarly, for the inverse Fourier transform of $f(p)$, we obtain:

$$\langle p | x \rangle = \frac{1}{\sqrt{2\pi}} \cdot e^{-ipx}. \tag{9.28}$$

A function can be expressed using the delta function as

$$f(x) = \int dx_0 \cdot \delta(x - x_0) f(x_0).$$

On the other hand, using the bra-ket notation,

$$f(x) = \langle x | f \rangle = \langle x | \cdot | f \rangle = \langle x | \int dx_0 \cdot |x_0\rangle\langle x_0||f\rangle$$

$$= \int dx_0 \, \langle x | x_0 \rangle \, f(x_0).$$

Comparing these two expressions, we obtain the relationship

$$\delta(x - x_0) = \langle x | x_0 \rangle. \tag{9.29}$$

9.3 Supplementary Derivations in Electromagnetism

9.3.1 *Formula for the Inverse-square Potential*

Let us explicitly compute Eq. (3.2). The calculation proceeds as follows:

$$\begin{aligned}\left[\nabla\left(\frac{1}{r}\right)\right]_j &= \partial_j\left((x_l x_l)^{-1/2}\right) = -\frac{1}{2}(x_l x_l)^{-3/2}\cdot\partial_j(x_l x_l)\\ &= -\frac{1}{2}(x_l x_l)^{-3/2}\cdot\left(1\cdot x_j + x_j\cdot 1\right) = -x_j\,(x_l x_l)^{-3/2}\\ &= -x_j\cdot r^{-3}.\end{aligned}$$

The result for the j-th component can be written in vector notation as:

$$\left[\nabla\left(\frac{1}{r}\right)\right]_j = -\frac{x_j}{r^3}\quad,\quad\therefore\quad\nabla\left(\frac{1}{r}\right) = -\frac{\mathbf{r}}{r^3}.$$

This reproduces Eq. (3.1).

Next, by differentiating Eq. (3.1) again, and noting that $\partial_l(r_l) = 3$ [as per Eq. (2.11)], we have:

$$\begin{aligned}-\nabla^2\left(\frac{1}{r}\right) &= \nabla\cdot\left(\frac{\mathbf{r}}{r^3}\right) = \partial_l\left(\frac{r_l}{r^3}\right) = -3r^{-4}(\partial_l r)\cdot r_l + \frac{1}{r^3}(\partial_l r_l)\\ &= -\frac{3\cdot r_l}{r^4}\cdot\partial_l\left[(r_m r_m)^{1/2}\right] + \frac{3}{r^3}\\ &= -\frac{3\cdot r_l}{r^4}\cdot\frac{1}{2}(r_m r_m)^{-1/2}\,\partial_l\left[r_m r_m\right] + \frac{3}{r^3}\\ &= -\frac{3\cdot r_l}{r^4}\cdot\frac{1}{2r}\cdot 2r_l + \frac{3}{r^3} = -\frac{3\cdot r_l r_l}{r^5} + \frac{3}{r^3}\\ &= -3r^{-3} + 3r^{-3} = 0.\end{aligned}$$

Thus, the result is obtained.

9.3.2 *Supplementary Derivations for the Vector Potential*

Regarding (3.26), expressing the cross product in terms of the Eddington epsilon notation, we have:

$$\left[\nabla \times \left(\frac{\mathbf{j}}{\phi(\mathbf{r})}\right)\right]_j = \varepsilon_{jlm}\left[\partial_l\left(\frac{1}{\phi(\mathbf{r})}\right) j_m\right] = \left[\nabla\left(\frac{1}{\phi(\mathbf{r})}\right)\times \mathbf{j}\right]_j$$
$$= -\left[\mathbf{j}\times\nabla\left(\frac{1}{\phi(\mathbf{r})}\right)\right]_j.$$

For (3.30), we can write:

$$[\nabla(\nabla \times \mathbf{A})] = \partial_l \varepsilon_{lst}\partial_s A_t = \varepsilon_{lst}\cdot\partial_l\partial_s A_t.$$

Since $\partial_l\partial_s$ can be interchanged for indices l and s, as in the case of (2.26), it follows that:

$$[\nabla(\nabla \times \mathbf{A})] = \varepsilon_{lst}\cdot\partial_l\partial_s A_t = 0.$$

Thus, the expression is identically zero.

Regarding (3.31), we have:

$$\begin{aligned}
[\nabla \times (\nabla \times \mathbf{A})]_j &= \varepsilon_{jlm}\cdot\partial_l(\nabla\times\mathbf{A})_m\\
&= \varepsilon_{jlm}\partial_l\cdot\varepsilon_{mst}\partial_s A_t = \varepsilon_{jlm}\varepsilon_{mst}\cdot\partial_l\partial_s A_t\\
&= \varepsilon_{mjl}\varepsilon_{mst}\cdot\partial_l\partial_s A_t = \left(\delta_{js}\delta_{lt}-\delta_{jt}\delta_{ls}\right)\cdot\partial_l\partial_s A_t\\
&= \partial_l\partial_j A_l - \partial_l\partial_l A_j = \partial_j(\nabla\cdot\mathbf{A}) - \nabla^2 A_j\\
&= [\nabla(\nabla\cdot\mathbf{A})]_j - \left[\nabla^2\mathbf{A}\right]_j.
\end{aligned}$$

Thus, from (2.21):

$$\nabla \times (\nabla \times \mathbf{A}) = \nabla(\nabla\cdot\mathbf{A}) - \nabla^2\mathbf{A}.$$

This result is derived as shown.

9.3.3 *Electromotive Force*

Electric current refers to the "movement of charge", and charges move under the influence of electric forces. The magnitude of the current density $\mathbf{j}(\mathbf{r})$ is, therefore, proportional to the magnitude of the electric field $\mathbf{E}$ at that location. This proportional relationship is observed as:

$$\mathbf{j} = \sigma\mathbf{E},$$

which is known as **Ohm's Law**, with σ referred to as the electrical conductivity.

When current flows through a closed circuit c, the line integral of the current density $\mathbf{j}(\mathbf{r})$ along the circuit,

$$\frac{1}{\sigma}\oint_c d\mathbf{l}\cdot\mathbf{j}(\mathbf{r}) = V_{\text{ex}},$$

is referred to as the **electromotive force**. If a larger current flows in the circuit, it implies that the electromotive force V_{ex} driving it is larger.

Ohm's Law describes the electric field $\mathbf{E}$ driving the motion of charge q as $\mathbf{E} = \mathbf{j}/\sigma$. However, the field describing current in a closed circuit is not a static electric field but a different type of electric field, as can be inferred from the following. Taking the line integral of Ohm's Law:

$$\frac{1}{\sigma}\oint_C d\mathbf{l}\cdot\mathbf{j}(\mathbf{r}) = \oint_c d\mathbf{l}\cdot\mathbf{E}(\mathbf{r}) = \oint_S d\mathbf{S}\cdot(\nabla\times\mathbf{E}(\mathbf{r}))\,.$$

The static electric field $\mathbf{E}(\mathbf{r})$ is given by $\mathbf{E}(\mathbf{r}) = -\nabla\phi(\mathbf{r})$, where $\phi(\mathbf{r})$ is the electrostatic potential. As a consequence, $\nabla\times\mathbf{E}(\mathbf{r}) = 0$ (Sect. 3.2.2), which implies that the right-hand side of the above equation becomes zero. Thus, the phenomenon of "current flowing along a closed circuit" cannot occur in a purely static electric field. Therefore, when current flows along a closed circuit, it must be described as:

$$\mathbf{j}(\mathbf{r},t) = \sigma\left[\mathbf{E}(\mathbf{r}) + \mathbf{E}_{\text{ex}}(\mathbf{r},t)\right],$$

where $\mathbf{E}(\mathbf{r})$ represents the static electric field, and $\mathbf{E}_{\text{ex}}(\mathbf{r},t)$ is an additional dynamic electric field component. Consequently:

$$\frac{1}{\sigma}\oint_c d\mathbf{l}\cdot\mathbf{j}(\mathbf{r},t) = \oint_c d\mathbf{l}\cdot\left[\mathbf{E}(\mathbf{r}) + \mathbf{E}_{\text{ex}}(\mathbf{r},t)\right] = \oint_c d\mathbf{l}\cdot\mathbf{E}_{\text{ex}}(\mathbf{r},t) = V_{\text{ex}} \neq 0.$$

This logic explains how a non-zero current can be sustained in the circuit.

Dynamic electric fields can thus be expressed as:

$$\mathbf{E}(\mathbf{r},t) = \mathbf{E}(\mathbf{r}) + \mathbf{E}_{\text{ex}}(\mathbf{r},t)\,,$$

i.e., "the sum of the static electric field and the electromotive force field".

The electromotive force is given by:

$$V_{\text{ex}} = \oint_c d\mathbf{l}\cdot\mathbf{E}_{\text{ex}}(\mathbf{r},t) = \frac{1}{\sigma}\oint_c d\mathbf{l}\cdot\mathbf{j}(\mathbf{r},t)\,.$$

From this line integral, we define the potential

$$\phi_{\text{ex}}(\mathbf{r}) = \int_L^{\mathbf{r}} d\mathbf{l}\cdot\mathbf{E}(\mathbf{r},t)\,,$$

leading to:

$$\nabla \phi_{\text{ex}} (\mathbf{r}, t) = \frac{1}{\sigma} \cdot \mathbf{j} (\mathbf{r}, t),$$

which establishes a depiction where "the vector field of current" is regarded as "the gradient of an underlying scalar field (potential)."[14] The potential difference $\delta \phi_{\text{ex}} (\mathbf{r}, t)$ thus:

$$\delta \phi_{\text{ex}} (\mathbf{r}, t) \propto \frac{1}{\sigma} \cdot \mathbf{j} (\mathbf{r}, t),$$

gives rise to the current. The voltage v, measured with respect to a reference potential ϕ_0 (e.g., ground), is:

$$v = \phi_{\text{ex}} (\mathbf{r}, t) - \phi_0.$$

This yields:

$$v \propto \mathbf{j} (\mathbf{r}, t),$$

which is the familiar form of Ohm's law.

9.3.4 Electric Field, Magnetic Field, Electric Flux Density, and Magnetic Flux Density

In Coulomb's law (3.15) and the Biot-Savart law (3.23), the reasoning for factoring out $1/4\pi$ is explained in the footnote of Eq. (3.22). However, Coulomb's law defines ε_0 in the denominator, while the Biot-Savart law places μ_0 in the numerator. These constants are referred to as **dielectric constant** and **magnetic permeability**, respectively, with the subscript "0" indicating "the value in a vacuum". The definition of ε_0, which places a constant in the denominator, might feel counterintuitive for beginners. To build an intuitive understanding of what these constants signify, consider the following reasoning: Coulomb's law can be interpreted as $\rho \propto \varepsilon_0 \cdot E$, while the Biot-Savart law reads as $B \propto \mu_0 \cdot j$. For the magnetic field, we recall that "an electric current generates a magnetic field", and for the electric field, "applying an electric field induces charge". These causal relationships guide the convention of associating the response coefficients, dielectric constant and magnetic permeability,

[14] This requires caution. Recall the discussion in Sect. 3.1.6: potentials defined only at single points in $\mathbf{r}$ ensure that closed-loop integrals vanish. For $\phi_{\text{ex}} (\mathbf{r})$, however, the closed-loop integral is non-zero. The variable $\mathbf{r}$ for $\phi_{\text{ex}} (\mathbf{r})$ is confined to points along the wire, avoiding the stricter criteria expected of potentials in free space, such as path-independence.

in the respective equations. Accordingly, ε_0 appears in the denominator and μ_0 in the numerator by this convention.

In the text, $\mathbf{B}(\mathbf{r})$ is referred to as the magnetic field, but strictly speaking, it is called the **magnetic flux density**. Additionally, in Sect. 3.3.1, the concept of electric flux density, $\mathbf{D}(\mathbf{r})$, was introduced. For beginners, this definition of quantities involving "flux" might seem a bit confusing, as one might wonder why the terms electric field and magnetic field alone do not suffice. For Coulomb's law with a general dielectric constant ε,

$$E = \frac{1}{4\pi\varepsilon} \cdot \frac{q}{r^2},$$

it can be rewritten as

$$D = \varepsilon \cdot E = \frac{q}{4\pi r^2}.$$

This allows us to interpret $D = \varepsilon \cdot E$ as the electric flux density, with the idea that "q electric flux lines emerge per unit sphere $4\pi r^2$".

By adopting what is now considered a less conventional approach [5] of introducing the Biot–Savart law as a counterpart to Coulomb's law using a hypothetical magnetic monopole q_m, magnetic flux density can be introduced quite naturally. In this formulation, the force between magnetic monopoles is expressed as

$$F_m = \frac{1}{4\pi\mu} \cdot \frac{q_m q_m'}{r^2} = q_m \cdot H.$$

Here, μ appears in the denominator, analogous to ε, addressing the discomfort mentioned in the first paragraph. In this formulation, H corresponds to the force field akin to the electric field (as part of the EH analogy [6]), and it is H that is referred to as the magnetic field. Accordingly,

$$B = \mu \cdot H = \frac{q_m}{4\pi r^2},$$

allowing $B = \mu \cdot H$ to be interpreted as magnetic flux density. This can be visualized as "q_m magnetic flux lines emerging per unit sphere $4\pi r^2$". Since magnetic monopoles are typically assumed not to exist, standard textbooks do not adopt the approach of treating magnetic forces analogously to electric forces. As a result, the reasoning behind calling B the magnetic flux density is less intuitively understood in conventional treatments.

At the elementary outline level of this book, the relationships $\mathbf{B} = \mu_0 \cdot \mathbf{H}$ and $\mathbf{D} = \varepsilon_0 \cdot \mathbf{E}$ might give the impression that they are essentially the same, merely proportional. However, the deeper understanding of why the four fields (E, D, H, B) are treated independently in the formulation of Maxwell's equations is not widely comprehended, even among researchers, including the author [6].

According to the reference [6], "It is insufficient to think that (**E**, **B**) pertains to vacuum, while (**D**, **H**) pertains to material medium. The correct interpretation is that (**E**, **H**) are force fields, while (**D**, **B**) are source fields". The text goes further to state, "The former force fields are 1-forms (first-order covariant tensors, covectors), whereas the latter source fields are 2-forms (second-order antisymmetric covariant tensors), belonging to different hierarchies as quantities. When electromagnetism was first formulated, **E** was represented as equipotential surfaces, and **D** as electric flux tubes, reflecting their geometric meanings. However, these geometric characteristics have been lost in the modern description of electromagnetism using vector calculus". While beginners can treat this as something to note in passing for now, it is worth mentioning that claiming to fully understand electromagnetism at this level may invite criticism from experts in the field. This is especially important to bear in mind when engaging with discussions on this subject [6].

9.4 Notes on Analytical Mechanics

9.4.1 Omission of Canonical Formulation

Although this book does not delve into the topic, the foundational courses underpinning the electronic state calculations specialized by my research group cover subjects such as second quantization and phonons. Second quantization serves as an essential "language" for handling many-body electronic theory. Additionally, the theory of phonons[15] is fundamental for understanding electronic properties such as thermal conductivity and superconductivity. The entry point to this "mountain" is analytical mechanics described in terms of the Poisson bracket formalism. From here, one ascends toward the "Heisenberg formulation of quantum mechanics" (**matrix mechanics**). With the knowledge covered up to Sect. 4 in this book, you can directly proceed in this direction.[16] Phonons and second quantization are discussed through a process known as **canonical quantization**, which is addressed within the framework of matrix mechanics (Fig. 9.4).

In a mechanics course aimed at bridging to quantum mechanics, the curriculum naturally includes not only the transition from the Hamiltonian-Jacobi formulation to Schrodinger's wave mechanics but also the connection from the Poisson bracket formulation to Heisenberg's matrix mechanics. This book omits the latter part.

In many modern courses, wave mechanics is introduced first, followed by a transformation of perspective regarding whether the state evolves over time or the operators do. This approach transitions from wave mechanics (the former perspective) to matrix mechanics (the latter perspective). The resulting Heisenberg equation of motion is then compared with the equation of motion expressed using

[15] The quantized treatment of lattice vibrations within materials.

[16] This is discussed in another book authored by me [7].

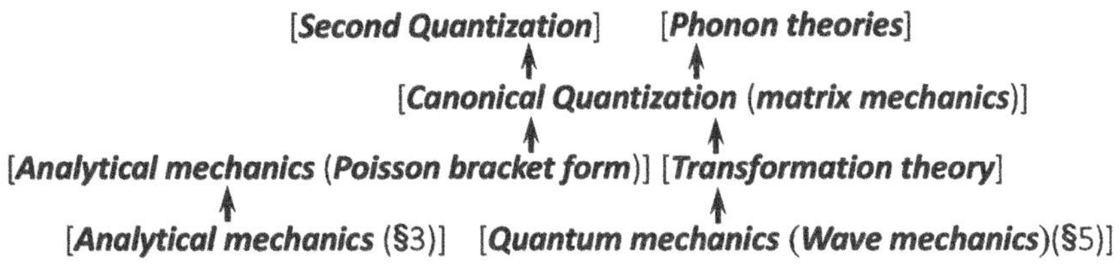

Fig. 9.4 The content centered on canonical quantization omitted in this book

Poisson brackets in classical mechanics to explain the procedure of canonical quantization.[17] However, the historical development of matrix mechanics originated from the perplexing experimental facts of the time, leading through the organization of knowledge such as Ritz's combination principle and the Bohr-Sommerfeld quantization rules, and was born out of insights that were somewhat more grounded and intuitive [8]. Studying these topics provides valuable insight into understanding "how our predecessors constructed the theories".

9.4.2 Designing a Curriculum for Analytical Mechanics

This book is designed with a "minimal climbing route" that provides a clear pathway to more advanced topics, but analytical mechanics, with its long history, offers a variety of "climbing routes", ranging from traditional curricula[11] to highly mathematically formalized ones[12]. Traditional curricula often begin with the **principle of virtual work**, building towards Lagrangian mechanics. This approach uses concepts like "equilibrium with inertial forces" and "constraint forces do no work" to eliminate constraint conditions from the system of equations, leading to the formulation of analytical mechanics [11]. Such formulations, like thermodynamics, present a challenge: "While it makes sense when explained by a teacher, can one reproduce the same reasoning independently?" It requires a highly careful "flow of discussion", where, for instance, "why certain variables are chosen as independent is not always self-evident", making it somewhat daunting for beginners. This book adopts a concise introductory style that avoids the profound depths of analytical mechanics[13].

In analytical mechanics, discussions have been carried out by reformulating problems into various forms, such as the Lagrangian formalism, the Hamiltonian formalism, the Hamilton-Jacobi formalism, and the Poisson bracket formalism.[18] Among these, the Lagrangian formalism is highly useful for "solving practical

[17] The origin of this approach in educational materials can be traced back to Pauli's work[10] as noted in other literature[9].

[18] The Poisson bracket formalism is omitted in this book.

problems", and traditionally serves as a major application field for **mechanical dynamics**, one of the so-called "Four Dynamics,"[19] commonly studied in mechanical engineering departments.

In courses aimed at bridging to quantum mechanics, two pathways are typically emphasized: one leading from the Hamilton-Jacobi formalism to Schrodinger's wave mechanics, and the other leading from the Poisson bracket formalism to Heisenberg's matrix mechanics. As in this course, reaching the Hamilton-Jacobi formalism provides a smooth and intuitive pathway to wave mechanics. However, this approach to introduction is generally followed only in physics departments, and it is rarely discussed in the curricula of engineering disciplines that apply quantum mechanics.[20] This book omits the "Poisson bracket…" pathway, although this approach seems to be more commonly included even in engineering curricula.[21]

One possible reason why the Hamilton-Jacobi formalism is less frequently taught today is that the traditional approach presented in classical textbooks tends to be complex and difficult to follow. Traditional curricula often derive the Hamilton-Jacobi equation by addressing topics such as the principal function of canonical transformations, first integrals, and cyclic coordinates[11]. These themes are developed primarily from a mathematical perspective, focusing on the "solvability of differential equations". For beginners, this approach can be confusing, as it obscures the question of "what perspective is being adopted and what the equation is intended to describe". This lack of clarity may have led to the gradual omission of the Hamilton-Jacobi formalism from many courses. This book, however, avoids such a traditional path and instead adopts a logical flow based on the perspective of "how the time evolution of the action function S at its endpoints is described" (Sect. 4.5), as suggested by another approach [14].

9.5 Topics Related to Relativity

9.5.1 Analytical Mechanics of Fields

Let us consider the Lagrangian introduced in classical particle mechanics, $L\left(\mathbf{q}_j(t), \dot{\mathbf{q}}_j(t)\right)$. Here, j serves as an index labeling particles. By explicitly denoting the vector components with the index l, we write:

$$L\left(q_j^{(l)}(t), \dot{q}_j^{(l)}(t)\right).$$

[19] The "Four Dynamics" refers to mechanical dynamics, thermodynamics, strength of materials, and fluid dynamics.

[20] There have often been cases where individuals claiming, "Well, it was taught in my department!", are actually referring to the Hamiltonian formalism rather than the Hamilton-Jacobi formalism.

[21] It appears to be included for the purpose of explaining the uncertainty principle.

From this, we expand the treatment to a continuous field

$$q_j^{(l)}(t) \to \phi_l(\mathbf{r}, t),$$

where the discrete particle index j is replaced by the position coordinate $\mathbf{r}$. Through this substitution, the canonical formalism, extended to describe the dynamics of the field $\phi_l(\mathbf{r}, t)$, can be developed.

The variable $q_j^{(l)}(t)$ depended only on t, involving a single variable. However, for $\phi_l(\mathbf{r}, t)$, it involves four variables: $\mathbf{r}$ and t. Correspondingly, the differentiated derivatives also include four types, generalizing the Lagrangian to incorporate spatial derivatives and spatial dependencies:

$$L(q_l(t), \dot{q}_l(t)) \to \mathcal{L}(\phi_l(\mathbf{r}, t), \partial_t \phi_l(\mathbf{r}, t), \nabla\phi_l(\mathbf{r}, t)).$$

Correspondingly, the action integral

$$S = \int dt \cdot L\left(q_j^{(l)}(t), \dot{q}_j^{(l)}(t)\right),$$

is extended to

$$S = \int d^4x \cdot \mathcal{L}\left(\phi_l\left(x_\mu\right), \partial_\mu \phi_l\left(x_\mu\right)\right),$$

where the four-component integration accounts for time and space (x_1, x_2, x_3, t), compactly represented as "4-dimensional spacetime (x_1, x_2, x_3, x_4)" (Sect. 6.2.2).

In this case,

$$\begin{aligned} S &= \int d^4x \cdot \mathcal{L}\left(\phi_l\left(x_\mu\right), \partial_\mu \phi_l\left(x_\mu\right)\right) \\ &= \int dt \int d^3r \cdot \mathcal{L}(\phi_l(\mathbf{r}, t), \partial_t \phi_l(\mathbf{r}, t), \nabla\phi_l(\mathbf{r}, t)) \\ &= \int dt \cdot L, \end{aligned}$$

so by integrating out only the spatial dependence first,

$$L = \int d^3x \cdot \mathcal{L}(\phi_l(\mathbf{r}, t), \partial_t \phi_l(\mathbf{r}, t), \nabla\phi_l(\mathbf{r}, t)),$$

this portion becomes the Lagrangian, and the script $\mathcal{L}$ corresponds to the Lagrangian density.

The Euler-Lagrange equation is derived from the variation of the action integral:

$$\delta S = \int d^4x \left\{ \frac{\partial \mathcal{L}}{\partial \phi_l} \cdot \delta\phi_l + \frac{\partial \mathcal{L}}{\partial \partial_\mu \phi_l} \cdot \delta\partial_\mu \phi_l \right\},$$

as we recall from (Sect. 4.3.2). By following the same logic as the "derivation through partial integration" performed in classical particle mechanics, we obtain:

$$\partial_\mu \frac{\partial \mathcal{L}}{\partial \left(\partial_\mu \phi_l\right)} - \frac{\partial \mathcal{L}}{\partial \phi_l} = 0,$$

which serves as the "equation governing the minimization of the action."[22] Through this generalized form, the **Lagrangian for fields** is defined as "a function that provides the governing equations for fields using the above Euler-Lagrange equation".

For example, the Lagrangian that yields the wave equation:

$$\partial_t^2 \phi(t,x) = c^2 \cdot \partial_x^2 \phi(t,x),$$

is given by:

$$\mathcal{L} = \frac{\rho}{2} \cdot [\partial_t \phi(t,x)]^2 - \frac{\lambda}{2} \cdot [\partial_x \phi(t,x)]^2,$$

where it can be verified that $c = \lambda/\rho$.

The Hamiltonian formalism for fields is constructed as follows. Following from the formalism in particle system analytical mechanics:

$$p^{(l)} = \frac{\partial L}{\partial \left(\dot{q}^{(l)}\right)} = \frac{\partial L}{\partial \left(\partial_4 q^{(l)}\right)}, \quad , \quad H = \sum_l p^{(l)} \cdot \left(\partial_4 q^{(l)}\right) - L,$$

one introduces the momentum for fields:

$$\pi_l = \frac{\partial \mathcal{L}}{\partial \left(\partial_4 \phi_l\right)},$$

and constructs the field Hamiltonian via a Legendre transformation:

$$\mathcal{H} = \sum_l \pi_l \cdot (\partial_4 \phi_l) - \mathcal{L}.$$

In non-relativistic problems, it is customary to begin by defining the Hamiltonian. However, in fields such as particle physics, the discussion typically starts with

[22] Note that the index μ in the first term is summed over, as per Einstein's summation convention.

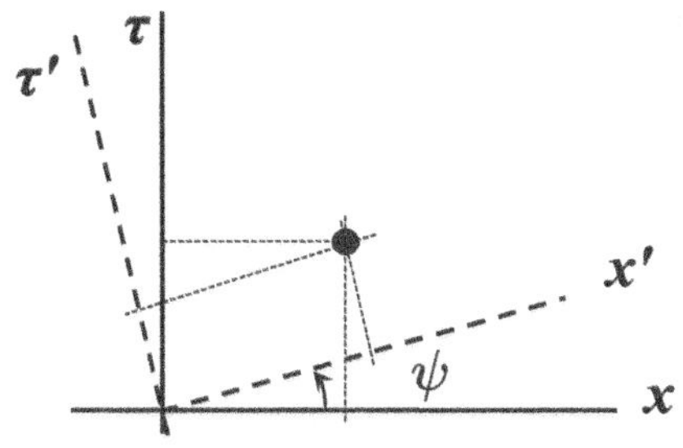

Fig. 9.5 Isometric transformation. The components of the same point are reinterpreted in a different coordinate system while preserving the distance from the origin

the Lagrangian formalism rather than the Hamiltonian formalism. This approach is used because the target is treated using the analytical mechanics of fields, where the Lagrangian formalism provides better symmetry properties in relativistic contexts. Specifically, it treats time and space equivalently, ensuring relativistic invariance[15].

9.5.2 Lorentz Transformation

Equation (6.11) can be expressed using the imaginary time $\tau = ict$ as[23]

$$x^2 + y^2 + z^2 + \tau^2 = x'^2 + y'^2 + z'^2 + \tau'^2.$$

Transformations of coordinate systems that satisfy this constraint lead to "transformations where time and space mix", as shown in Eq. (9.31), and this will be derived below.

This is an isometric transformation in four-dimensional space, so the coordinates before and after the transformation (those with primes) are related by trigonometric functions as shown in Fig. 9.5 (Sect. 9.2.4).

Let us now align the y-axis and z-axis while considering a scenario where the x-axis moves uniformly with velocity v relative to the fixed x'-axis (set to $x' = 0$). In this case, only x and τ are involved in the isometric transformation, which, based on Fig. 9.5, gives:

$$x = x' \cdot \cos\psi - \tau' \cdot \sin\psi \quad , \quad \tau = x' \cdot \sin\psi + \tau' \cdot \cos\psi. \tag{9.30}$$

By substituting $x' = 0$, we get:

$$x = -\tau' \cdot \sin\psi, \quad , \quad \tau = -\tau' \cdot \cos\psi, \quad , \quad \therefore \quad \tan\psi = -\frac{x}{\tau} = -\frac{vt}{ict} = i\frac{v}{c}.$$

[23] Here, x, etc., are used instead of dx, assuming the origin is at zero.

From this, we calculate:

$$\cos^2\psi = \frac{1}{1+\tan^2\psi} = \frac{1}{1+(iv/c)^2} = \frac{1}{1-(v/c)^2},$$

$$\sin^2\psi = 1-\cos^2\psi = \frac{\tan^2\psi}{1+\tan^2\psi} = \frac{-(v/c)^2}{1-(v/c)^2}.$$

Taking the positive root:

$$\cos\psi = \frac{1}{\sqrt{1-(v/c)^2}}, \quad , \quad \sin\psi = \frac{i\,(v/c)}{\sqrt{1-(v/c)^2}}.$$

Substituting these into Eq. (9.30), we have:

$$\begin{aligned} x &= x' \cdot \frac{1}{\sqrt{1-(v/c)^2}} - \tau' \cdot \frac{i\,(v/c)}{\sqrt{1-(v/c)^2}} \\ &= \frac{x' - i\,(v/c)\,\tau'}{\sqrt{1-(v/c)^2}} = \frac{x'+vt'}{\sqrt{1-(v/c)^2}}, \\ \tau &= x' \cdot \frac{i\,(v/c)}{\sqrt{1-(v/c)^2}} + \tau' \cdot \frac{1}{\sqrt{1-(v/c)^2}} \\ &= \frac{ix' \cdot (v/c) + \tau'}{\sqrt{1-(v/c)^2}}. \end{aligned}$$

Rewriting in terms of t and x:

$$x = \frac{x'+vt'}{\sqrt{1-(v/c)^2}}, \quad , \quad t = \frac{x' \cdot \left(v/c^2\right) + t'}{\sqrt{1-(v/c)^2}}. \tag{9.31}$$

This transformation is known as the **Lorentz transformation**. In the limit $c \to \infty$, this reduces to:

$$x = x' + vt', \quad , \quad t = t',$$

which corresponds to the Galilean transformation, matching our classical intuition. By treating the speed of light c as finite, the Lorentz transformation (Eq. 9.31) reveals that "time and space hybridize equivalently".

Dividing the first equation of (9.31) by the second, we get:

$$\frac{x}{t} = \frac{x' + Vt'}{t' + x'\left(V/c^2\right)}$$

$$\therefore \quad v = \frac{x'\left(1 + V \cdot t'/x'\right)}{t'\left[1 + (x'/t')\left(V/c^2\right)\right]}$$

$$= v'\frac{1 + V/v'}{1 + v'\left(V/c^2\right)} = \frac{v' + V}{1 + v'\left(V/c^2\right)},$$

which yields (6.10).

9.5.3 The Principle of Invariant Light Speed

The principle of invariant light speed, when presented abruptly, can be quite difficult for beginners to accept, as it often leads to the question, "Why does this concept emerge?" Historically, during the period when special relativity was being developed, topics such as "Lorentz transformations in electromagnetism" and the "Michelson-Morley experiment" were prominent[16]. Understanding this principle becomes much more intuitive when these contexts are considered together.

The Michelson-Morley experiment aimed to determine whether "if light propagates through a medium called ether, the speed of light should appear different when measured in directions perpendicular or parallel to the Earth's orbital motion". The experiment sought to detect this difference. However, the result revealed no such discrepancy, leading to the conclusion that the intuition of light speed being subject to a Galilean transformation such as $c \pm v$ could not be validated.

"The Lorentz transformation in electromagnetism" can be introduced through the following reasoning. Around a steady straight current, a magnetic field is generated surrounding the current. Meanwhile, a straight stationary charge produces a radial electric field. A steady current is essentially formed by charges moving at a constant velocity. Consequently, an observer moving with the same velocity as the current should perceive the magnetic field as a radial electric field. According to the principle of relativity, which asserts that the laws of physics must remain invariant between inertial reference frames, the observation of a magnetic field appearing as an electric field seems, at first glance, to contradict this principle. However, this issue is resolved by the fact that the space and time coordinates in Maxwell's equations, which govern electromagnetic fields, transform as in Eq. (9.31). This transformation ensures that such phenomena, where a magnetic field appears as an electric field, are consistent with the principle of relativity [17]. The Lorentz transformation was discovered within such a context. However, as seen in the previous section, the Lorentz transformation can also be derived by postulating the invariance of the speed of light. This connection likely strengthened the conviction that the invariance of the speed of light holds as a fundamental principle.

It is worth noting that in Landau's textbook [14], the principle of invariance of the speed of light is argued without relying on the above historical background. The reasoning presented there is as follows:[24] "In Newtonian mechanics, which aligns with the intuitive velocity addition rule (Galilean transformation), there is an implicit assumption that interactions propagate instantaneously. Empirically, interactions are known to propagate at finite speeds. Assuming that there is a maximum propagation speed, denoted as c, no motion should exceed this value. If such motion existed, it would allow interactions to propagate faster than c, violating the premise that c is the maximum. If the principle of relativity is to be upheld, the governing laws must be identical in all reference frames, and c must also be identical in all frames. Consequently, no speed can exceed this value c" (End of quote).

9.5.4 Derivation of the Spin-Magnetic Field Coupling Term

Using the notation for the commutator $[A, B] := AB - BA$ and the anticommutator $\{A, B\} := AB + BA$, the Pauli matrices in (6.23) satisfy the following properties:

$$\left\{\sigma_j, \sigma_k\right\} = 2\delta_{jk} \quad , \quad \left[\sigma_j, \sigma_k\right] = 2i \cdot \varepsilon_{jkl} \cdot \sigma_l .$$

Thus,

$$\begin{aligned} 2\sigma_j\sigma_k &= \left\{\sigma_j, \sigma_k\right\} + \left[\sigma_j, \sigma_k\right] = 2\delta_{jk} + 2i \cdot \varepsilon_{jkl} \cdot \sigma_l \\ \therefore \quad \sigma_j\sigma_k &= \delta_{jk} + i \cdot \varepsilon_{jkl} \cdot \sigma_l , \end{aligned} \tag{9.32}$$

and (6.30) is obtained.

In the component form of the Pauli matrices involved in (6.29), applying the above relations, we have:

$$\begin{aligned} \sigma_l\sigma_m \quad \cdot & \; (p_l - q\, A_l)\,(p_m - q\, A_m) \\ &= (\delta_{lm} + i \cdot \varepsilon_{lms} \cdot \sigma_s)\,(p_l - q\, A_l)\,(p_m - q\, A_m) \\ &= (p_l - q\, A_l)\,(p_l - q\, A_l) \\ & \quad + i \cdot \varepsilon_{lms} \cdot \sigma_s \left(p_l p_m + q^2 A_l A_m - q\, p_l A_m - q\, p_m A_l\right) . \end{aligned}$$

---

[24] However, one might wonder if this explanation is readily accessible to beginners outside the field.

In the second term, $p_l p_m$ and $A_l A_m$ combined with ε_{lms} do not contribute because $\mathbf{a} \times \mathbf{a} = 0$. Thus, we have:

$$\begin{aligned}\sigma_l \sigma_m \quad &\cdot \quad (p_l - q\ A_l)\,(p_m - q\ A_m) \\ &= (\mathbf{p} - q\mathbf{A})^2 - i \cdot q \cdot (\sigma_s \varepsilon_{slm}\, p_l A_m + \sigma_s \varepsilon_{slm} A_l\, p_m) \\ &= (\mathbf{p} - q\mathbf{A})^2 - iq \cdot \boldsymbol{\sigma} \cdot (\mathbf{p} \times \mathbf{A} + \mathbf{A} \times \mathbf{p})\,.\end{aligned}$$

Using vector notation:

$$\boldsymbol{\sigma} \cdot (\mathbf{p} - q\mathbf{A})\, \boldsymbol{\sigma} \cdot (\mathbf{p} - q\mathbf{A}) = (\mathbf{p} - q\mathbf{A})^2 - iq\ \boldsymbol{\sigma} \cdot (\mathbf{p} \times \mathbf{A} + \mathbf{A} \times \mathbf{p}) \qquad (9.33)$$

Substituting $\mathbf{p} = (\hbar/i)\,\nabla$ into the above equation, the second term becomes:

$$-\,q\hbar\ \sigma_l\ \varepsilon_{lmn}\ \partial_m\,(A_n \cdot \psi) \sim \sigma_l\ \varepsilon_{lmn}\ \psi\ (\partial_m A_n) + \sigma_l\ \varepsilon_{lmn}\ A_n\ (\partial_m \psi)\,.$$

Since $\mathbf{p} \sim \nabla$ acts as a differential operator, it is essential to account for its operation on ψ as shown above. The second term on the right-hand side cancels with the third term in (9.33) after interchanging the order of the cross product. Thus:

$$\boldsymbol{\sigma} \cdot (\mathbf{p} - q\mathbf{A})\, \boldsymbol{\sigma} \cdot (\mathbf{p} - q\mathbf{A}) = (\mathbf{p} - q\mathbf{A})^2 - q\hbar\ \boldsymbol{\sigma} \cdot (\nabla \times \mathbf{A}) \qquad (9.34)$$

and (6.31) is obtained.

9.5.5 Eigenvalue Shift by Diagonal Terms

Consider the eigenvalue problem for a matrix that includes diagonal terms:

$$C = a \cdot I + B$$

with the corresponding eigenvalue equation:

$$C \cdot \mathbf{v}^{(C)} = \lambda^{(C)} \cdot \mathbf{v}^{(C)}.$$

Now, let the eigenvalues and eigenvectors of the matrix B be:

$$B \cdot \mathbf{v}^{(B)} = \lambda^{(B)} \cdot \mathbf{v}^{(B)}.$$

When the matrix C acts on the eigenvector $\mathbf{v}^{(B)}$, we have:

$$\begin{aligned}C \cdot \mathbf{v}^{(B)} &= (a \cdot I + B) \cdot \mathbf{v}^{(B)} = a \cdot \mathbf{v}^{(B)} + \lambda^{(B)} \cdot \mathbf{v}^{(B)} \\ &= \left(a + \lambda^{(B)}\right) \cdot \mathbf{v}^{(B)}.\end{aligned}$$

This shows that the eigenvalues and eigenvectors of the matrix C are:

$$\lambda^{(C)} = \left(a + \lambda^{(B)}\right) \quad , \quad \mathbf{v}^{(C)} = \mathbf{v}^{(B)}.$$

9.5.6 Eigenstates of the Pauli Matrices

Let us determine the eigenvectors of the Pauli matrices introduced in (6.23):

$$\sigma_x = \begin{pmatrix} 0 & 1 \\ 1 & 0 \end{pmatrix} \quad , \quad \sigma_y = \begin{pmatrix} 0 & -i \\ i & 0 \end{pmatrix} \quad , \quad \sigma_z = \begin{pmatrix} 1 & 0 \\ 0 & -1 \end{pmatrix}.$$

For σ_z, the equation to solve is:

$$\begin{pmatrix} 1-\lambda & 0 \\ 0 & -1-\lambda \end{pmatrix} \begin{pmatrix} x_1 \\ x_2 \end{pmatrix} = 0,$$

and the characteristic equation:

$$(1-\lambda)(-1-\lambda) = 0$$

yields the eigenvalues $\lambda = 1, -1$. For each eigenvalue, substituting into the equation gives:

$$\begin{pmatrix} 0 & 0 \\ 0 & -2 \end{pmatrix} \begin{pmatrix} x_1 \\ x_2 \end{pmatrix}_{\lambda=1} = 0 \quad , \quad \begin{pmatrix} 2 & 0 \\ 0 & 0 \end{pmatrix} \begin{pmatrix} x_1 \\ x_2 \end{pmatrix}_{\lambda=-1} = 0.$$

This gives $x_1 = 0$ or $x_2 = 0$, respectively. Therefore:

$$\begin{pmatrix} x_1 \\ x_2 \end{pmatrix}_{\lambda=1} = \begin{pmatrix} 1 \\ 0 \end{pmatrix} \quad , \quad \begin{pmatrix} x_1 \\ x_2 \end{pmatrix}_{\lambda=-1} = \begin{pmatrix} 0 \\ 1 \end{pmatrix},$$

namely:

$$|\psi\rangle^{(z)}_{\lambda=1} = |1\rangle \quad , \quad |\psi\rangle^{(z)}_{\lambda=-1} = |0\rangle$$

are the eigenstates.

Similarly, for σ_x and σ_y, the equations to solve are:

$$\begin{pmatrix} 0-\lambda & 1 \\ 1 & 0-\lambda \end{pmatrix} \begin{pmatrix} x_1 \\ x_2 \end{pmatrix}_{(\alpha=x)} = 0 \quad , \quad \begin{pmatrix} 0-\lambda & -i \\ i & 0-\lambda \end{pmatrix} \begin{pmatrix} x_1 \\ x_2 \end{pmatrix}_{(\alpha=y)} = 0,$$

and in both cases, the characteristic equation is $\lambda^2 = 1$. For σ_x, substituting λ into the equations gives:

$$\begin{pmatrix} -1 & 1 \\ 1 & -1 \end{pmatrix} \begin{pmatrix} x_1 \\ x_2 \end{pmatrix}_{(\alpha=x)(\lambda=1)} = 0 \quad , \quad \begin{pmatrix} 1 & 1 \\ 1 & 1 \end{pmatrix} \begin{pmatrix} x_1 \\ x_2 \end{pmatrix}_{(\alpha=x)(\lambda=-1)} = 0.$$

From these, we find:

$$\begin{pmatrix} x_1 \\ x_2 \end{pmatrix}_{(\alpha=x)(\lambda=1)} = \begin{pmatrix} 1 \\ 1 \end{pmatrix} \quad , \quad \begin{pmatrix} x_1 \\ x_2 \end{pmatrix}_{(\alpha=x)(\lambda=-1)} = \begin{pmatrix} 1 \\ -1 \end{pmatrix},$$

namely:

$$|\psi\rangle^{(x)}_{\lambda=1} = \frac{1}{\sqrt{2}}(|0\rangle + |1\rangle) \quad , \quad |\psi\rangle^{(x)}_{\lambda=-1} = \frac{1}{\sqrt{2}}(|0\rangle - |1\rangle),$$

which are the eigenstates.

For σ_y, the equations to solve are:

$$\begin{pmatrix} -1 & -i \\ i & -1 \end{pmatrix} \begin{pmatrix} x_1 \\ x_2 \end{pmatrix}_{(\alpha=y)(\lambda=1)} = 0 \quad , \quad \begin{pmatrix} 1 & -i \\ i & 1 \end{pmatrix} \begin{pmatrix} x_1 \\ x_2 \end{pmatrix}_{(\alpha=y)(\lambda=-1)} = 0.$$

From these, we find:

$$\begin{pmatrix} x_1 \\ x_2 \end{pmatrix}_{(\alpha=y)(\lambda=1)} = \begin{pmatrix} 1 \\ i \end{pmatrix} \quad , \quad \begin{pmatrix} x_1 \\ x_2 \end{pmatrix}_{(\alpha=y)(\lambda=-1)} = \begin{pmatrix} 1 \\ -i \end{pmatrix}.$$

Namely:

$$|\psi\rangle^{(y)}_{\lambda=1} = \frac{1}{\sqrt{2}}(|0\rangle + i \cdot |1\rangle) \quad , \quad |\psi\rangle^{(y)}_{\lambda=-1} = \frac{1}{\sqrt{2}}(|0\rangle - i \cdot |1\rangle),$$

which are the eigenstates.

9.5.7 Eigenvectors in Arbitrary Directions

For the matrix

$$R = \begin{pmatrix} \cos\theta & \sin\theta \cdot e^{-i\phi} \\ \sin\theta \cdot e^{i\phi} & -\cos\theta \end{pmatrix},$$

the eigenvalue equation $R \cdot \mathbf{x} = \lambda \cdot \mathbf{x}$ is

$$\begin{pmatrix} \cos\theta - \lambda & \sin\theta \cdot e^{-i\phi} \\ \sin\theta \cdot e^{i\phi} & -\cos\theta - \lambda \end{pmatrix} \begin{pmatrix} x_1 \\ x_2 \end{pmatrix} = 0, \tag{9.35}$$

and the corresponding characteristic equation is

$$(\lambda - \cos\theta)(\lambda + \cos\theta) - \sin^2\theta = 0 \quad , \quad \therefore \quad \lambda = \pm 1 = \lambda_{\pm}.$$

For λ_+, the equation becomes

$$\begin{pmatrix} \cos\theta - 1 & \sin\theta \cdot e^{-i\phi} \\ \sin\theta \cdot e^{i\phi} & -\cos\theta - 1 \end{pmatrix} \begin{pmatrix} x_1 \\ x_2 \end{pmatrix} = 0.$$

From the first row of the eigenvalue Eq. (9.35),

$$(\cos\theta - 1) \cdot x_1 + \sin\theta \cdot e^{-i\phi} \cdot x_2 = 0.$$

Using the trigonometric identities:

$$\cos\theta = \cos\frac{\theta}{2}\cos\frac{\theta}{2} - \sin\frac{\theta}{2}\sin\frac{\theta}{2}, \quad , \quad \sin\theta = \sin\frac{\theta}{2}\cos\frac{\theta}{2} + \cos\frac{\theta}{2}\sin\frac{\theta}{2},$$

$$1 = \cos\frac{\theta}{2}\cos\frac{\theta}{2} + \sin\frac{\theta}{2}\sin\frac{\theta}{2},$$

the equation simplifies to

$$-2\sin\frac{\theta}{2}\sin\frac{\theta}{2} \cdot x_1 + 2\sin\frac{\theta}{2}\cos\frac{\theta}{2} \cdot e^{-i\phi} \cdot x_2 = 0$$

$$\therefore \quad -\sin\frac{\theta}{2} \cdot x_1 + \cos\frac{\theta}{2} \cdot e^{-i\phi} \cdot x_2 = 0.$$

Thus, the eigenvector is given by:

$$\begin{pmatrix} x_1 \\ x_2 \end{pmatrix} = \begin{pmatrix} \cos\frac{\theta}{2} \\ e^{i\phi} \cdot \sin\frac{\theta}{2} \end{pmatrix}.$$

9.6 Supplementary Notes on Field Transformations and Spin

9.6.1 Commutation Relations of Angular Momentum Operators

The angular momentum operator introduced in (7.5),

$$\hat{\mathbf{l}} = (-i)\ \mathbf{r} \times \nabla,$$

can be expressed in component form as

$$\hat{l}_\gamma = (-i)\ \varepsilon_{\gamma st} \cdot r_s \partial_t, \tag{9.36}$$

where the indices γ, s, and t represent spatial coordinates $(x,\ y,\ z)$. Recall that summation over the indices s and t is implied on the right-hand side. We now derive the commutation relation of the angular momentum operator:

$$\hat{l}_\alpha \hat{l}_\beta - \hat{l}_\beta \hat{l}_\alpha = i \cdot \varepsilon_{\alpha\beta\gamma} \hat{l}_\gamma, \tag{9.37}$$

as below.

Acting the first term on the left-hand side of (9.37) on a function $\psi(\mathbf{r})$, we have:

$$\begin{aligned}
\hat{l}_\alpha \hat{l}_\beta \cdot \psi &= -\varepsilon_{\alpha st} r_s \partial_t \cdot \varepsilon_{\beta mn} r_m \partial_n \cdot \psi \\
&= -\varepsilon_{\alpha st} \varepsilon_{\beta mn} \cdot r_s \cdot \partial_t \left[r_m \left(\partial_n \psi \right) \right] \\
&= -\varepsilon_{\alpha st} \varepsilon_{\beta mn} \cdot r_s \left[\left(\partial_t r_m \right) \left(\partial_n \psi \right) + r_m \left(\partial_t \partial_n \psi \right) \right] \\
&= -\varepsilon_{\alpha st} \varepsilon_{\beta mn} \cdot r_s \left(\partial_t r_m \right) \left(\partial_n \psi \right) - \varepsilon_{\alpha st} \varepsilon_{\beta mn} \cdot r_s r_m \left(\partial_t \partial_n \psi \right)
\end{aligned} \tag{9.38}$$

The second term on the right-hand side can be evaluated by systematically replacing the dummy indices as $(s, t) \to \left(m', n'\right) \to (m, n)$:

$$\begin{aligned}
(\text{2nd term}) &= -\varepsilon_{\alpha st} \varepsilon_{\beta mn} \cdot r_s r_m \partial_t \partial_n \psi \\
&= -\varepsilon_{\alpha m'n'} \varepsilon_{\beta s't'} \cdot r_{m'} r_{s'} \partial_{n'} \partial_{t'} \psi \\
&= -\varepsilon_{\beta st} \varepsilon_{\alpha mn} \cdot r_m r_s \partial_n \partial_t \psi \\
&= -\varepsilon_{\beta st} \varepsilon_{\alpha mn} \cdot r_s r_m \partial_t \partial_n \psi
\end{aligned}$$

This result matches the expression obtained by swapping α and β in the original equation. Thus, the terms symmetric in α and β, originating from $\hat{l}_\alpha \hat{l}_\beta \psi$ and $\hat{l}_\beta \hat{l}_\alpha \psi$, cancel each other.

Thus, in the evaluation of (9.38), it suffices to focus solely on the first term. This can be evaluated as:[25]

$$
\begin{aligned}
\hat{l}_\alpha \hat{l}_\beta \psi - \hat{l}_\beta \hat{l}_\alpha \psi &= -\varepsilon_{\alpha st}\varepsilon_{\beta mn} \cdot r_s (\partial_t r_m)(\partial_n \psi) + \varepsilon_{\beta st}\varepsilon_{\alpha mn} \cdot r_s (\partial_t r_m)(\partial_n \psi) \\
&= \left(-\varepsilon_{\alpha st}\varepsilon_{\beta mn} + \varepsilon_{\beta st}\varepsilon_{\alpha mn}\right) r_s \cdot \delta_{tm} \cdot \partial_n \psi \\
&= \left(-\varepsilon_{\alpha sm}\varepsilon_{\beta mn} + \varepsilon_{\beta sm}\varepsilon_{\alpha mn}\right) r_s \partial_n \psi \\
&= \left(-\varepsilon_{m\alpha s}\varepsilon_{mn\beta} + \varepsilon_{m\beta s}\varepsilon_{mn\alpha}\right) r_s \partial_n \psi \\
&= \left\{-\left(\delta_{\alpha n}\delta_{s\beta} - \delta_{\alpha\beta}\delta_{ns}\right) + \left(\delta_{\beta n}\delta_{s\alpha} - \delta_{\beta\alpha}\delta_{sn}\right)\right\} r_s \partial_n \psi \\
&= \left\{-\delta_{\alpha n}\delta_{s\beta} + \delta_{\beta n}\delta_{s\alpha}\right\} r_s \partial_n \psi \\
&= \left(r_\alpha \partial_\beta - r_\beta \partial_\alpha\right) \psi \\
&= \left(\delta_{s\alpha}\delta_{t\beta} - \delta_{s\beta}\delta_{t\alpha}\right) r_s \partial_t \psi \\
&= \varepsilon_{\gamma\alpha\beta} \cdot \varepsilon_{\gamma st} r_s \partial_t \psi.
\end{aligned} \tag{9.39}
$$

From the definition of the angular momentum operator in (9.36):

$$
\varepsilon_{st\gamma} r_s \partial_t = i \hat{l}_\gamma,
$$

we substitute this into (9.39) to obtain the following relation:

$$
\left(\hat{l}_\alpha \hat{l}_\beta - \hat{l}_\beta \hat{l}_\alpha\right) \psi = i \varepsilon_{\alpha\beta\gamma} \hat{l}_\gamma. \tag{9.40}
$$

9.6.2 *Angular Momentum Algebra and the Dimensions of Rotation Representations*

In general, for an algebra satisfying:

$$
\left[\hat{l}_\alpha, \hat{l}_\beta\right] = i \cdot \varepsilon_{\alpha\beta\gamma} \cdot \hat{l}_\gamma, \tag{9.41}
$$

we define:

$$
\hat{l}_\pm := \frac{1}{\sqrt{2}}\left(\hat{l}_x \pm i \cdot \hat{l}_y\right), \tag{9.42}
$$

[25] In the transition from the first to the second line, $(\partial_t r_m) = \delta_{tm}$ is applied. In the transition from the third to the fourth line, index cycling of $\varepsilon_{\alpha\beta\gamma}$ is performed, and (2.21) is applied in the transition from the fourth to the fifth line. In the transition from the fifth to the sixth line, symmetry $\delta_{ij} = \delta_{ji}$ is used.

and it can be verified that:

$$\left[\hat{l}_z, \hat{l}_\pm\right] = \pm\hat{l}_\pm, \quad , \quad \left[\hat{l}_+, \hat{l}_-\right] = \hat{l}_z, \tag{9.43}$$

hold true.[26]

Let us expressed the eigenvalues and eigenstates of $\hat{l}_z$ as:

$$\hat{l}_z|m\rangle = m|m\rangle.$$

Applying (9.43) to $\hat{l}_z\hat{l}_\pm$ acting on this $|m\rangle$, we have:

$$\begin{aligned}\hat{l}_z\hat{l}_\pm|m\rangle &= \left(\hat{l}_\pm\hat{l}_z \pm \hat{l}_\pm\right)|m\rangle = \hat{l}_\pm\hat{l}_z|m\rangle \pm \hat{l}_\pm|m\rangle \\ &= \hat{l}_\pm m|m\rangle \pm \hat{l}_\pm|m\rangle = (m \pm 1)\,\hat{l}_\pm|m\rangle.\end{aligned}$$

Rewriting, this can be expressed as:

$$\hat{l}_z \cdot \underline{\hat{l}_\pm|m\rangle} = (m \pm 1) \cdot \underline{\hat{l}_\pm|m\rangle}.$$

Thus, the state $\underline{\hat{l}_\pm|m\rangle}$ is one that possesses an eigenvalue of $(m \pm 1)$ with respect to $\hat{l}_z$. That is:

$$\hat{l}_\pm|m\rangle \sim |m \pm 1\rangle.$$

In other words, the operator $\hat{l}_\pm$ acts on the state $|m\rangle$ to increase or decrease m, and is referred to as a **ladder operator**.

Now, let us define the maximum eigenvalue M such that:

$$\hat{l}_+|M\rangle = 0,$$

indicating that "no higher state with increased m exists". At this stage, M has not been specified as an integer or real number, but it will later be shown to be either an integer or a half-integer. When the lowering operator acts on this state, assume that:

$$\hat{l}_-|M\rangle = N_M \cdot |M - 1\rangle,$$

where N_M is a coefficient. Taking the conjugate of both sides gives:

$$\langle M|\hat{l}_+ = N_M^* \cdot \langle M - 1|.$$

[26] These relations can be directly derived by substituting (9.42) into the left-hand side and applying (9.41).

Multiplying both sides yields:

$$\langle M|\hat{l}_+\hat{l}_-|M\rangle = |N_M|^2 \langle M-1|M-1\rangle = |N_M|^2 .$$

Thus, we find:

$$\begin{aligned}|N_M|^2 &= \langle M|\hat{l}_+\hat{l}_-|M\rangle \\ &= \langle M|\left(\hat{l}_-\hat{l}_+ + \hat{l}_z\right)|M\rangle = \langle M|\hat{l}_z|M\rangle = M, \\ \therefore \quad N_M &= \sqrt{M}.\end{aligned}$$

This determines the value of N_M.

Next, let us evaluate $\hat{l}_+|M-1\rangle$:

$$\begin{aligned}\hat{l}_+|M-1\rangle &= \hat{l}_+\left(\frac{1}{N_M}\hat{l}_-|M\rangle\right) \\ &= \frac{1}{N_M}\hat{l}_+\hat{l}_-|M\rangle = \frac{1}{N_M}\left(\hat{l}_-\hat{l}_+ + \hat{l}_z\right)|M\rangle \\ &= \frac{1}{N_M}\hat{l}_z|M\rangle = \frac{M}{N_M}|M\rangle \\ &= \frac{M}{\sqrt{M}}|M\rangle = \sqrt{M}|M\rangle = N_M|M\rangle.\end{aligned}$$

Summarizing the results:

$$\begin{aligned}\hat{l}_-|M\rangle &= N_M|M-1\rangle, \\ \hat{l}_+|M-1\rangle &= N_M|M\rangle, \quad N_M = \sqrt{M}.\end{aligned} \tag{9.44}$$

This relationship can be extended by decrementing M as follows:

$$\begin{aligned}\hat{l}_-|M-1\rangle &= N_{M-1}|M-2\rangle, \\ \hat{l}_+|M-2\rangle &= N_{M-1}|M-1\rangle \\ \hat{l}_-|M-2\rangle &= N_{M-2}|M-3\rangle, \\ \hat{l}_+|M-3\rangle &= N_{M-2}|M-2\rangle \\ &\cdots \\ \hat{l}_-|M-k\rangle &= N_{M-k}|M-k-1\rangle, \\ \hat{l}_+|M-k-1\rangle &= N_{M-k}|M-k\rangle.\end{aligned}$$

For the final result:

$$\hat{l}_-|M-k\rangle = N_{M-k}|M-k-1\rangle,$$

and its conjugate:

$$\langle M-k|\hat{l}_+ = N_{M-k}^*\langle M-k-1|,$$

taking the product of both sides yields:

$$|N_{M-k}|^2 = \langle M-k|\hat{l}_+\hat{l}_-|M-k\rangle = \langle M-k|\left(\hat{l}_-\hat{l}_+ + \hat{l}_z\right)|M-k\rangle.$$

Here, since:

$$\hat{l}_+|M-k\rangle = N_{M-k+1}|M-k+1\rangle,$$

we obtain:

$$\begin{aligned}
|N_{M-k}|^2 &= \langle M-k|\hat{l}_-\hat{l}_+|M-k\rangle + \langle M-k|\hat{l}_z|M-k\rangle \\
&= |N_{M-k+1}|^2 + (M-k)\,, \\
\therefore \quad |N_{M-k}|^2 - |N_{M-k+1}|^2 &= (M-k)\,.
\end{aligned}$$

By aligning with $|N_M|^2 = M$ and incrementing k step by step, we write:

$$\begin{aligned}
|N_M|^2 &= M \\
|N_{M-1}|^2 - |N_M|^2 &= (M-1) \\
|N_{M-2}|^2 - |N_{M-1}|^2 &= (M-2) \\
&\cdots \\
|N_{M-k}|^2 - |N_{M-k+1}|^2 &= (M-k)\,.
\end{aligned}$$

Summing these equations side by side yields:

$$\begin{aligned}
|N_{M-k}|^2 &= M \times (k+1) - (1+2+\cdots+k) \\
&= M \times (k+1) - \frac{1}{2}\cdot k\,(k+1) \\
&= (k+1)\left(M - \frac{k}{2}\right).
\end{aligned}$$

Therefore, we obtain:

$$|N_{M-k}|^2 = \frac{(2M-k)(k+1)}{2}.$$

Revisiting the scenario where the maximum eigenvalue for $\hat{l}_z|m\rangle = m|m\rangle$ is M, and using:

$$\hat{l}_-|M\rangle = N_M|M-1\rangle,$$

we sequentially decrement the eigenvalue by constructing $|M-k\rangle$. In this case:

$$|N_{M-k}|^2 = \frac{(2M-k)(k+1)}{2},$$

indicating that at $k = 2M$, the coefficient N_{M-k} becomes zero, and further states cannot be constructed. Thus, organizing the independent eigenstates of $\hat{l}_z$ yields:

$$\begin{aligned} &\{|M\rangle, |M-1\rangle, \cdots, |M-2M\rangle\} \\ &= \{|M\rangle, |M-1\rangle, \cdots, |-M\rangle\}. \end{aligned}$$

The total number of eigenfunctions is $(2M+1)$, indicating a natural number. This requirement implies that the initially assumed maximum eigenvalue M can be either an integer or a half-integer.

9.6.3 Representation Matrices for Spinor Fields

For the two-dimensional field with $N = (2M+1) = 2$ $(M = 1/2)$, the two states can be written as:

$$|\pm M\rangle = \{|+1/2\rangle, |-1/2\rangle\} = \{|\uparrow\rangle, |\downarrow\rangle\}.$$

Using the rules derived earlier:

$$\begin{aligned} \hat{l}_z|M\rangle &= M|M\rangle, \\ \hat{l}_-|M\rangle &= \sqrt{M}|M-1\rangle, \\ \hat{l}_+|M-1\rangle &= \sqrt{M}|M\rangle, \end{aligned}$$

we construct the representation matrix from (7.11):

$$L_{ij}^{(\alpha)} = \begin{pmatrix} \langle\uparrow|\hat{l}_\alpha|\uparrow\rangle & \langle\uparrow|\hat{l}_\alpha|\downarrow\rangle \\ \langle\downarrow|\hat{l}_\alpha|\uparrow\rangle & \langle\downarrow|\hat{l}_\alpha|\downarrow\rangle \end{pmatrix}.$$

The components are:

$$L_{ij}^{(z)} = \frac{1}{2}\begin{pmatrix} \langle\uparrow|\uparrow\rangle & 0 \\ 0 & -\langle\downarrow|\downarrow\rangle \end{pmatrix} = \frac{1}{2}\begin{pmatrix} 1 & 0 \\ 0 & -1 \end{pmatrix},$$

$$L_{ij}^{(+)} = \begin{pmatrix} 0 & \sqrt{2^{-1}}\langle\uparrow|\uparrow\rangle \\ 0 & \sqrt{2^{-1}}\langle\downarrow|\uparrow\rangle \end{pmatrix} = \frac{1}{\sqrt{2}}\begin{pmatrix} 0 & 1 \\ 0 & 0 \end{pmatrix},$$

$$L_{ij}^{(-)} = \begin{pmatrix} \sqrt{2^{-1}}\langle\uparrow|\downarrow\rangle & 0 \\ \sqrt{2^{-1}}\langle\downarrow|\downarrow\rangle & 0 \end{pmatrix} = \frac{1}{\sqrt{2}}\begin{pmatrix} 0 & 0 \\ 1 & 0 \end{pmatrix}.$$

By solving (9.42) for $\hat{l}_\pm$, we find:

$$2\hat{l}_x = \sqrt{2}\left(\hat{l}_+ + \hat{l}_-\right), \quad 2i\cdot\hat{l}_y = \sqrt{2}\left(\hat{l}_+ - \hat{l}_-\right).$$

By inserting the basis functions on both sides, we have:

$$2L^{(x)} = \sqrt{2}\left(L^{(+)} + L^{(-)}\right) = \begin{pmatrix} 0 & 1 \\ 1 & 0 \end{pmatrix},$$

$$2i\cdot L^{(y)} = \sqrt{2}\left(L^{(+)} - L^{(-)}\right) = \begin{pmatrix} 0 & 1 \\ -1 & 0 \end{pmatrix}.$$

Thus, we obtain:

$$L_{ij}^{(z)} = \frac{1}{2}\begin{pmatrix} 1 & 0 \\ 0 & -1 \end{pmatrix}, \quad L_{ij}^{(x)} = \frac{1}{2}\begin{pmatrix} 0 & 1 \\ 1 & 0 \end{pmatrix}, \quad L_{ij}^{(y)} = \frac{1}{2}\begin{pmatrix} 0 & -i \\ i & 0 \end{pmatrix}.$$

9.6.4 Azimuthal Angle of the Spinor

From (7.15) and (7.16),

$$\hat{U}(\mathbf{n})\cdot|\psi_\mathbf{n}\rangle = |\psi_\mathbf{n}\rangle \quad , \quad \hat{U}(\mathbf{n})_{jk} = \left[n_x\cdot\sigma_x + n_y\cdot\sigma_y + n_z\cdot\sigma_z\right],$$

the eigenvalue equation can be written as

$$
\begin{aligned}
|\psi_{\mathbf{n}}\rangle &= \left(n_x\sigma_x + n_y\sigma_y + n_z\sigma_z\right)|\psi_{\mathbf{n}}\rangle \\
\therefore \quad \sigma_x|\psi_{\mathbf{n}}\rangle &= \left(n_x\sigma_x^2 + n_y\sigma_x\sigma_y + n_z\sigma_x\sigma_z\right)|\psi_{\mathbf{n}}\rangle \\
&= \left(n_x\cdot I + i\cdot n_y\sigma_z - i\cdot n_z\sigma_y\right)|\psi_{\mathbf{n}}\rangle \\
\therefore \quad \langle\psi_{\mathbf{n}}|\sigma_x|\psi_{\mathbf{n}}\rangle &= n_x + i\cdot n_y\langle\psi_{\mathbf{n}}|\sigma_z|\psi_{\mathbf{n}}\rangle \\
&\quad - i\cdot n_z\langle\psi_{\mathbf{n}}|\sigma_y|\psi_{\mathbf{n}}\rangle,
\end{aligned}
\tag{9.45}
$$

where the following relations were used:

$$
\sigma_x^2 = I \quad , \quad \sigma_i\sigma_j = i\cdot\varepsilon_{ijk}\cdot\sigma_k \quad (i\neq j) \quad , \quad \langle\psi_{\mathbf{n}}|\psi_{\mathbf{n}}\rangle = 1. \tag{9.46}
$$

By similarly applying σ_y and σ_z and evaluating as was done with σ_x, we introduce the notation

$$
u_\alpha := \langle\psi_{\mathbf{n}}|\sigma_\alpha|\psi_{\mathbf{n}}\rangle,
$$

and derive the following:

$$
\begin{aligned}
u_x &= n_x + i\cdot n_y u_z - i\cdot n_z u_y, \\
u_y &= -i\cdot n_x u_z + n_y + i\cdot n_z u_x, \\
u_z &= +i\cdot n_x u_y - i\cdot n_y u_x + n_z.
\end{aligned}
$$

These equations can be organized into the form:

$$
\begin{pmatrix} 1 & iu_z & -iu_y \\ -iu_z & 1 & iu_x \\ iu_y & -iu_x & 1 \end{pmatrix} \cdot \begin{pmatrix} n_x \\ n_y \\ n_z \end{pmatrix} = \begin{pmatrix} u_x \\ u_y \\ u_z \end{pmatrix},
$$

which can be solved as:

$$
\begin{pmatrix} n_x \\ n_y \\ n_z \end{pmatrix} = \begin{pmatrix} 1 & iu_z & -iu_y \\ -iu_z & 1 & iu_x \\ iu_y & -iu_x & 1 \end{pmatrix}^{-1} \cdot \begin{pmatrix} u_x \\ u_y \\ u_z \end{pmatrix}.
$$

The determinant D associated with the inverse matrix $A^{-1} = \mathrm{Cof}[A]/D$ is given by[27]

$$
\begin{aligned}
D &= 1 - u_x^2 - iu_z\left(-iu_z + u_x u_y\right) - iu_y\left(-iu_y - u_x u_z\right) \\
&= 1 - \left(u_x^2 + u_y^2 + u_z^2\right).
\end{aligned}
$$

[27] Refer to (2.72) for details about inverse matrices.

Using this determinant, the inverse matrix is evaluated as

$$\begin{pmatrix} 1 & iu_z & -iu_y \\ -iu_z & 1 & iu_x \\ iu_y & -iu_x & 1 \end{pmatrix}^{-1} = D^{-1} \cdot \begin{pmatrix} 1-u_x^2 & -iu_z + u_x u_y & -u_x u_z + iu_y \\ * & * & * \\ * & * & * \end{pmatrix}.$$

For example, evaluating n_x gives:

$$\begin{aligned} n_x &= D^{-1}\left\{\left(1-u_x^2\right)\cdot u_x + \left(-iu_z + u_x u_y\right)\cdot u_y + \left(-u_x u_z + iu_y\right)\cdot u_z\right\} \\ &= D^{-1}u_x\left\{1-u_x^2-u_y^2-u_z^2\right\} \\ &= u_x. \end{aligned} \tag{9.47}$$

Similarly, evaluating n_y and n_z yields:

$$\begin{aligned} n_x &= u_x = \langle\psi_{\mathbf{n}}|\sigma_x|\psi_{\mathbf{n}}\rangle, \\ n_y &= u_y = \langle\psi_{\mathbf{n}}|\sigma_y|\psi_{\mathbf{n}}\rangle, \\ n_z &= u_z = \langle\psi_{\mathbf{n}}|\sigma_z|\psi_{\mathbf{n}}\rangle. \end{aligned} \tag{9.48}$$

The spinor state in (1.2),

$$|\psi^{(2)}_{(\theta,\phi)}\rangle = \cos\frac{\theta}{2}\cdot|0\rangle + e^{i\phi}\sin\frac{\theta}{2}\cdot|1\rangle = \begin{pmatrix} \cos\frac{\theta}{2} \\ e^{i\phi}\sin\frac{\theta}{2} \end{pmatrix},$$

can be used to determine $\mathbf{n}$ such that $\hat{U}(\mathbf{n})\cdot|\psi_{\mathbf{n}}\rangle = |\psi_{\mathbf{n}}\rangle$ by evaluating the right-hand side of (9.48) with $|\psi_{\mathbf{n}}\rangle = |\psi^{(2)}_{(\theta,\phi)}\rangle$. For example:

$$\begin{aligned} n_x &= \langle\psi^{(2)}_{(\theta,\phi)}|\sigma_x|\psi^{(2)}_{(\theta,\phi)}\rangle \\ &= \left(\cos(\theta/2)\ e^{-i\phi}\sin(\theta/2)\right)\begin{pmatrix} 0 & 1 \\ 1 & 0 \end{pmatrix}\begin{pmatrix} \cos(\theta/2) \\ e^{i\phi}\sin(\theta/2) \end{pmatrix} \\ &= \sin\theta\cos\phi. \end{aligned}$$

Similarly, one can evaluate:

$$\begin{aligned} n_x &= \langle\psi^{(2)}_{(\theta,\phi)}|\sigma_x|\psi^{(2)}_{(\theta,\phi)}\rangle = \sin\theta\cos\phi, \\ n_y &= \langle\psi^{(2)}_{(\theta,\phi)}|\sigma_y|\psi^{(2)}_{(\theta,\phi)}\rangle = \sin\theta\sin\phi, \\ n_z &= \langle\psi^{(2)}_{(\theta,\phi)}|\sigma_z|\psi^{(2)}_{(\theta,\phi)}\rangle = \cos\theta. \end{aligned}$$

This demonstrates that $\mathbf{n}$ corresponds to the coordinates $(\sin\theta\cos\phi, \sin\theta\sin\phi, \cos\theta)$.

References

1. Riley KF, Hobson MP, Bence SJ (2006) Mathematical methods for physics and engineering, 3rd edn. Cambridge University Press, Cambridge. ISBN: 978-0521679718
2. Jackson JD (1998) Classical electrodynamics. Wiley, New York. ISBN: 978-0471309321
3. Dirac PAM (1988) The principles of quantum mechanics, 4th edn. Oxford University Press, Oxford . ISBN: 978-0198520115
4. Needham T (2023) Visual complex analysis. Oxford Univercity Press, Oxford. ISBN: 978-0192868923
5. Murakami M (2013) Naruhodo Denjikigaku. Kaimeisha. ISBN: 978-4875253006
6. Sato F, Kitano M (2018) Shin SI tan-i to Denjikigaku. Iwanami. ISBN: 978-4000612616
7. Maezono R (2025) Ab initio quantum Monte Carlo tutorial: going beyond DFT. Springer (to be published)
8. Weinberg S (2015) Lectures on quantum mechanics, 2nd edn. Cambridge University Press, Cambridge. ISBN: 978-1107111660
9. Takabayashi T (2010) Ryoushiron no Hattenshi. Chikuma. ISBN: 978-4480093196
10. Pauli W, Achuthan P (1980) General principles of quantum mechanics Springer, Berlin (Translation). ISBN: 978-3540098423
11. Goldstein H, Poole CP, Safko JL (2013) Classical mechanics. Pearson, London. ISBN: 978-1292026558
12. Arnol'd VI, Vogtmann K et al (1989) Mathematical methods of classical mechanics, 2nd edn. Springer, Berlin (Translation). ISBN: 978-0387968902
13. Morifuji M (2005) Ryoushiha no dynamics. Yoshioka, Tokyo. ISBN: 978-4842703336
14. Landau LD, Lifshitz EM (1980) The classical theory of fields. Pergamon, Oxford. ISBN: 978-0080250724
15. Shimizu A (2004) Ryoushiron no Kiso (Science-sha, 2004). ISBN: 978-4781910628
16. Mansfield MM, O'Sullivan C, Understanding physics, 3rd edn. Wiley, New York. ISBN: 978-1119519508
17. Feynman RP, The Feynman lectures on physics, New Millennium edn. Basic Books, New York City. ISBN: 978-0465023820

Index

R. Maezono, *Essentials for Deeper Understanding of Quantum Computing*,
https://doi.org/10.1007/978-981-96-5646-2

The manufacturer's authorised representative in the EU is Springer Nature Customer Service Centre GmbH, Europaplatz 3, 69115 Heidelberg, Germany. If you have any concerns regarding our products, please contact ProductSafety@springernature.com

Printed and bound by CPI Group (UK) Ltd, Croydon, CR0 4YY
03/02/2026
02046970-0003